NATURE IN THE WEST

Also by Richard and Jacob Rabkin

FIRE ISLAND: THE WONDERS OF
A BARRIER BEACH

NATURE GUIDE TO
FLORIDA

NATURE
IN THE
WEST

A HANDBOOK OF HABITATS

Richard and Jacob Rabkin

Holt, Rinehart and Winston   New York

FOR JUDY

Library of Congress Cataloging in Publication Data
Rabkin, Richard, 1932–
Nature in the West.
1. Biotic communities—West (U.S.)   2. Habitat
(Ecology)—West (U.S.)   I. Rabkin, Jacob.
QH104.5.W4R3   574.5'0978   81-791
AACR2

ISBN Hardcover: 0-03-053671-5
ISBN Paperback: 0-03-053666-9

First Edition

Designer: Joy Chu
Printed in the United States of America
1 3 5 7 9 10 8 6 4 2

Contents

Color plates can be found on the following pages:

Introduction

Ecological studies and environmental political action have become quite popular, yet the available books on these subjects fall into two categories: field guides to the individual families of the animal and plant kingdoms (which would weigh a ton if, to pursue a broad interest in the total community of nature, you were to carry them all into the field) and tourist guides, which take you only up to the doorway, so to speak, of the places you wish to explore. To remedy this situation, we have tried to create a new category: a kind of observer's guide. This book is not a field manual to those things that naturalists collect and identify. It is not a travel guide to places tourists seek out. But it can be used by naturalists and tourists. It is a book about ecosystems. And its aim is to explain and illustrate such systems so that an observer can then immerse himself in the sights and sounds of a region.

In a manner of speaking, ecosystems are what you get by mixing the naturalist's things in the tourist's places and letting them cook for several millennia. Just as flour, yeast, and water are transformed in the baking into a product that does not retain any of the characteristics of the original ingredients, eons of time have transformed nature and brought about the life systems we see today. The story of those transformations is another element in this book.

Environmentalists try to preserve these ecosystems through political action. Yet no law will succeed in preserving something that people know only from a distance or do not understand at close range. It is our conviction that if habitats are to survive, a new kind of appreciation—sometimes called ecological consciousness—must be developed. This appreciation or consciousness is quite different from the tourist's curiosity about visiting new places or the naturalist's desire to collect and catalogue, yet it obviously includes both interests. Our goal in producing this book is to encourage the reader to become aware of the atmosphere of a particular wilderness setting, the relationships among the various plant and animal species, and the environmental context in which every individual member of a species exists in nature. Such an awareness will, we hope, enhance his appreciation of nature even as it expands his knowledge.

To achieve our goal, we have organized this book according to ecological zones, illustrating each of these zones as they occur in the American West. Zones, of course, are not discrete; they tend to overlap. For that reason, all of the flora and fauna that appear in each illustration will not always be found in the spot on which you are standing. Nonetheless, an awareness of zones leads to an understanding of habitats, and from there, it is only a short step toward ready identification of all of the life-forms that make up an ecosystem.

The illustrations are also not intended to be scientifically precise. This is not a scientific manual. They are intended to highlight features—markings and tracks, for example—that expedite recognition. But they are also intended to give as full a picture as possible of the whole ecosystem in a particular zone. For this reason, they include uncommon as well as common animals, nocturnal as well as diurnal, and they show seasonal birds and flowers that will not be seen if the visitor arrives out of season. The difficulty of placing so many plants and animals in one illustration often required that the scale vary from one plant to another. Nonetheless, at no point has science been ignored or appearances distorted.

—R.R. and J.R.

WILDLIFE AREAS

The eight states that are generally considered to make up the West are Montana, Idaho, Wyoming, Nevada, Utah, Colorado, Arizona, and New Mexico. Together they comprise more than 30 percent of the land area of the contiguous United States. Almost half of the land in these states belongs to the United States government. The national parks make up only a small part of these federal lands. The greatest portion consists of national forests and open grazing lands.

Naturalists, geologists, and ecologists examine and study this area in terms of its major geographic regions. Except for Nevada, the entire area from Canada to Mexico is dominated by the Rocky Mountains, a complex chain of numerous interlocking ranges with intermittent prairies, pastoral parklands, and emerald valleys. Meandering along the jagged line of its peaks and ridges is the Continental Divide, a line between two gigantic watersheds that drain the flow of water to the east into the Atlantic Ocean or the Gulf of Mexico, or to the west toward the Pacific Ocean and the Gulf of California. Surrounding this mountainous area are the vast grasslands on the east, picturesque deserts to the south, and a barren sagebrush basin to the west.

For some three or four billion years, the earth has been in the process of change. Forces of construction, such as earthquakes and volcanoes, have been competing with such destructive forces as erosion, gravity, and glaciers. Seas invade the land, mountains are thrown up by forces deep within the earth, hills are leveled into plains, continents break and drift apart, and seas retreat from the land. The scenery of the universe is constantly being reshaped and resculptured. All these processes are slow and, except in earthquake zones, hardly noticeable. But time is in ample supply in the universe. A thousand years of man's life on earth is merely a mo-

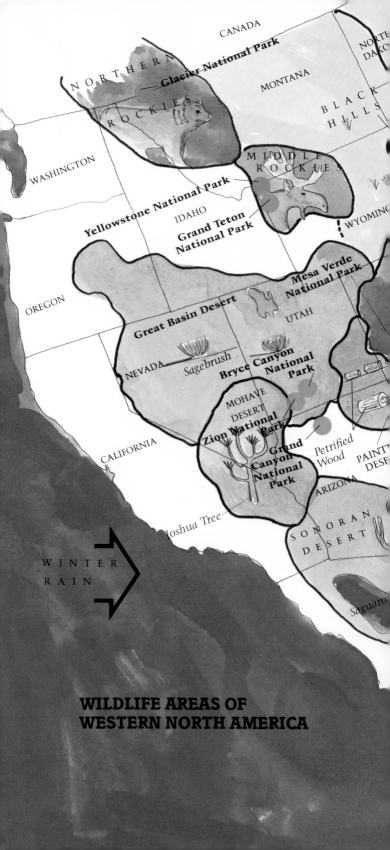

WILDLIFE AREAS OF WESTERN NORTH AMERICA

TALLGRASS PRAIRIE

UTH
KOTA

IXED-
GRASS

EBRASKA PLAINS

Future home of
Tallgrass Prairie
National Park

**Rocky Mountain
National Park**

KANSAS

OLORADO

SHORT
GRASS
STEPPE

OKLAHOMA

SOUTHERN
ROCKIES

EW MEXICO

TEXAS

**Big Bend
National Park**

SUMMER
RAIN

CHIHUAHUAN

DESERT

Lechuguilla

MEXICO

mentary flash across the colossal calendar of geological history. In terms of this billion-year history, the Rockies are relatively new. They had their ancestral beginnings some three hundred million years ago when the entire area was uplifted from shallow inland seas. This was followed by various cycles of invading seas and uplifting land. The vast sea retreated about seventy million years ago, and was followed by alternating periods of volcanic activity, erosion, and uplift. The final broad uplift of the Rocky Mountain area occurred about five to seven million years ago. This is recent geologic time, and it puts the Rockies in the "young" generation of mountains.

Mountains are made in any one of four major ways: faulting, folding, volcanic action, and doming. More than one of these forces often operates in the same area. Mountains also erode and decay in different ways. The result is an infinite variety of shapes and structures reflecting the operation of different combinations of these various forces. The geologist is sometimes hard put to disentangle and identify each of these.

The "faulted" mountain is often the most spectacular. It occurs along a weakness, or fault, in the earth's crust when the underground pressures become sufficiently great to force one whole mass of the earth's crustal rock to separate cleanly from another. On one side of the fracture, the rocks may slip or sink down. On the other side, the rocks may be thrust up and tilted, exposing an awesome wall of great blocks. A perfect example of a fault-block mountain is the jagged and angular Grand Tetons range, rising abruptly to a height of more than seven thousand feet above the Jackson Hole valley in Wyoming.

"Folded" mountains are created when internal pressures push sideways against great masses of rock. The earth's crust may be flexible enough to bend and buckle without any conspicuous cracking. The result will be a series of parallel ridges, which are to be found in many of the great mountain ranges, including the Rockies.

Volcanic activity is another mountain-building force. The raw material for all volcanoes is pockets of hot liquid, known as magma, brought up by enormous internal pressures from the earth's depths. Some volcanoes are explosive, spewing lava through a central vent that piles up in sufficient quantities to create the classic volcanic cone. The nonexplosive volcanoes pour lava out of craters or through long fissures in the earth, gradually building wide, gently sloping structures. Some volcanoes are active for centuries; inactive volcanoes may be extinct or merely dormant. There are many evidences of volcanic activity in the Rockies, including volcanic mountains and various forms of lava rock. In the Yellowstone area, water working its way down to the hot magma produces thousands of geysers, hot springs, and mudholes.

The fourth process of mountain building, doming, is the result of vulcanism. This process is to be distinguished from volcanism, which refers explicitly to the eruptive activity of volcanoes and the geological effects of volcanic lava, as magma is called when it reaches the surface. Volcanic activity may cause the rush of magma through a crack under existing layers of rock. Where the crack does not extend to the surface of the earth, the molten material collects in pockets and pushes up the overlying rocks without rupturing the surface. The result is a series of "domed" mountains, generally round or oval-shaped. Striking examples of this type of mountain-building are the Henry Mountains in Utah and the Black Hills of South Dakota.

The approach to the Rockies from the east is through the vast expanse of the plains country. Rainfall decreases as this open land rises beyond the west. As a result of this decrease, the tallgrass prairie of the moist lower slopes gives way to the mixed-grass plains of the middle ground, and then to the shortgrass steppe of the drier uplands.

The approach to the Rockies from the west is through a series of deserts. The most extensive of these is the Great Basin Desert, the so-called high desert, hemmed in on the east by the Rockies and on the west by the block mountains of the Sierra Nevada. Between these two barriers is an arid land of high plains and desert troughs. It is a lonely and windswept vastness dominated by sagebrush, a region of climatic extremes whose rivers and streams cannot reach the Pacific but must exhaust themselves in lakes, flats, and sinkholes.

Immediately to the south, in southern Nevada and southeastern California, the Great Basin gradually merges into the Mohave Desert. At the heart of this desert is Death Valley, the lowest point on the continent, consisting of two shallow basins surrounded by mountains, one of which rises to more than eleven thousand feet. The highest point in the West, Mt. Whitney, is only about a hundred miles away. The Mohave is one of the driest areas in the world; most of its limited precipitation comes during the cool winter months. The ground cover consists mainly of such shrubs as the creosote bush. In the higher elevations, the thick-branched Joshua tree, a giant yucca, can grow more than forty feet tall.

The third major desert division, a low desert compared to the Mohave, is the Sonoran, embracing the southeastern corner of California, southwestern Arizona, most of the Mexican state of Sonora, and most of Baja California. This, too, is dominated by a vast depressed basin, the Salton Sink, 250 miles long and almost 50 miles wide, which contains a lake whose surface is 200 feet below sea level. The Sonoran is rated as the most magnificent American desert, with a magical display of springtime wildflowers and a large variety of flowering cactus plants. Its most remarkable plant is the saguaro, which grows to a height of only twenty feet in seventy-five years and only fifty feet in two hundred.

To the southeast of the Sonoran Desert is Mexico's Chihuahuan Desert; this covers the broad plateau of north-central Mexico—an area about as extensive as the Great Basin Desert—with spurs extending into southern New Mexico and West Texas. Among the interesting features of this desert, all to be found in New Mexico, are tall dunes of dazzling white sand, extensive deposits of volcanic lava, and the remnant beaches of ancient lakes. Its vegetation is mainly small cacti, spiny shrubs, and spine-tipped yuccas. Its plant emblem is the agave known as lechuguilla, which sends up a ten-foot-high narrow cluster of small yellow leaves from the basal rosette of a rigid, sharply pointed plant.

Another product of upheaval activity is the desert in northeastern Arizona known as the Painted Desert. Here, because of the scarce vegetation, the skin-deep coloration of the rocks in the plateau country can be seen in all their glowing brilliance, their different hues depending upon the mineral or organic material originally buried in the eroded sediments of the plateau. In the eastern part of this plateau is the Petrified Forest. Here the cycle of erosion is more advanced, displaying remnants of an ancient forest that was buried by marshy sediments some two hundred million years ago and uplifted shortly before the raising of the Rockies.

A hundred miles east of the Rockies, but built by the same disturbances, lie the Black Hills of South Dakota, an isolated island of rock in the midst of the Great Plains. The uplift that caused these domal mountains occurred over a long period of time, and the erosion process began as soon as the surface started to rise. Thus, these mountains never reached their potential height, achieving an elevation of less than four thousand feet above the plains to the east. The loftiest dome is Harney Peak, a mass of pink granite about seven thousand feet above sea level. Mt. Rushmore, much lower, was chosen by the sculptor Gutzon Borglum for the presidential memorial because it consists of a single uniform mass of granite, a product of molten magma less vulnerable to the forces of erosion than sedimentary rock. The hills are covered by stands of pine and spruce that appear so dark against the arid plains that the name *Black Hills* seemed appropriate to the Native Americans and the early settlers alike.

LIFE ZONES

No two mountains are alike. They differ in size, composition, and climate. But they all have one thing in common—the higher the altitude, the lower the temperature. This is because high air, thinner than air at lower levels, absorbs less radiation from the sun. The result is that the average temperature on a mountain drops about three degrees Fahrenheit for every thousand-foot rise in altitude. This drop in temperature is roughly equivalent to the change that occurs on a six-hundred-mile journey north at sea level. Therefore, there should be some correlation between the changes in plant and animal life observed while climbing a mountain and the progression of major life-forms on a trip north to the Arctic.

The correlation, in certain respects, between latitudinal zones of climate, each with its distinct type of plant and animal life, and corresponding altitudinal zones of climate was developed into a system of life zones by C. Hart Merriam in the latter part of the nineteenth century. To modern-day scientists, however, every variable factor in the environment has some effect on the distribution of organisms. Rainfall, wind, fire, composition of the soil, and slope, as well as temperature, all contribute to the bewildering groupings of life-forms that develop in different environments. Other factors to be considered are the interactions of the organisms in each group, not only among themselves but with the physical environment as well.

The functioning unit resulting from the total interaction of plant and animal life (the biotic community) and its physical habitat (the abiotic conditions) is called an *ecological system*, or *ecosystem* for short. In this sense, the world is one large ecosystem. Ecologists, however, prefer to divide the world into numerous ecosystems, large and small. There is the ecosystem of a forest, a prairie, and a desert, and the ecosystem of a river, pond, or marsh. In each instance, the ecologist emphasizes the total environment to which an organism is adapted and the daily interactions that are constantly changing both the biotic community and the larger ecosystem.

Despite the preference of modern scientists for the ecosystem approach, Merriam's life-zone system survives as a useful way of discussing the effects of temperature on the distribution of different species of plant and animal life. Merriam divided North America into various major zones, ranging from tropical to arctic. In the same way, the ecosystem of the mountains can be considered as a series of temperature zones in which ascending altitude, rather than latitude, is the critical factor. However, these comparable life zones in the mountains are not necessarily stacked in neat horizontal sections. For example, a particular zone may extend a great deal farther up the warm southern slopes of a mountain—sometimes as much as two thousand feet— than on the north side. Nor are equivalent life zones found at the same altitude in different regions. Since it gets colder in the northern mountains, the boundaries are drawn at correspondingly lower elevations. Ponderosa-pine forests may be found at an elevation of four thousand feet in Montana and above eight thousand feet in Arizona, spruce-fir forests at six thousand feet in northern Idaho and at ten thousand feet in Colorado. Whatever its altitude, each zone can best be recognized by its typical plant and animal community.

From south to north, seven life zones are generally distinguished: Subtropical, Sonoran (or Lower Austral), Upper Sonoran (or Upper Austral), Foothills (or Transition), Montane (or Canadian), Subalpine (or Hudsonian), and Alpine (or Arctic). The Subtropical Zone is found only in small areas of the most southerly portion of the United States. None of it is located within the Rocky Mountain region.

The Sonoran Zone is the region of the American deserts. Here the dominant plants are the mesquite, a favorite firewood; the paloverde tree, whose tiny yellow

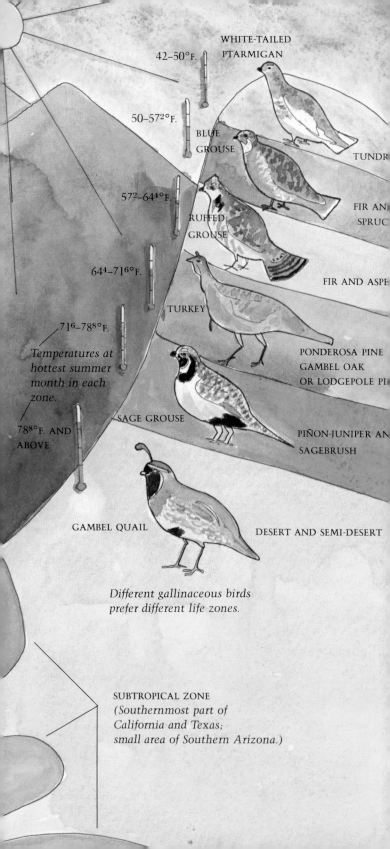

WHITE-TAILED
PTARMIGAN

42–50°F.

50–57²°F.

BLUE
GROUSE

TUNDR

57²–64⁴°F.

RUFFED
GROUSE

FIR AN
SPRUC

FIR AND ASPE

64⁴–71⁶°F.

TURKEY

PONDEROSA PINE
GAMBEL OAK
OR LODGEPOLE PI

71⁶–78⁸°F.

*Temperatures at
hottest summer
month in each
zone.*

PIÑON-JUNIPER AN
SAGEBRUSH

78⁸°F. AND
ABOVE

SAGE GROUSE

GAMBEL QUAIL

DESERT AND SEMI-DESERT

*Different gallinaceous birds
prefer different life zones.*

SUBTROPICAL ZONE
*(Southernmost part of
California and Texas;
small area of Southern Arizona.)*

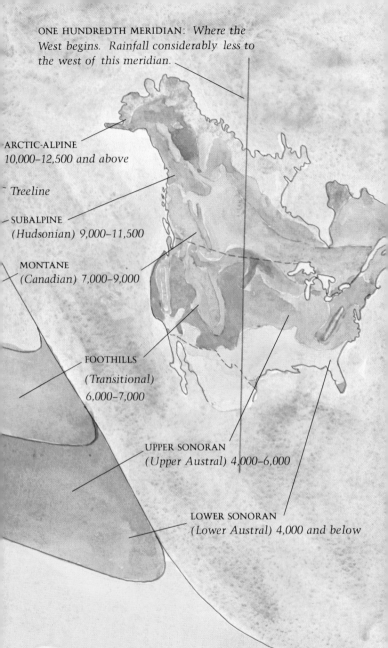

ONE HUNDREDTH MERIDIAN: *Where the West begins. Rainfall considerably less to the west of this meridian.*

ARCTIC-ALPINE
10,000–12,500 and above

Treeline

SUBALPINE
(Hudsonian) 9,000–11,500

MONTANE
(Canadian) 7,000–9,000

FOOTHILLS
(Transitional)
6,000–7,000

UPPER SONORAN
(Upper Austral) 4,000–6,000

LOWER SONORAN
(Lower Austral) 4,000 and below

LIFE ZONES IN WESTERN NORTH AMERICA

*Climbing a mountain is like traveling north.
The vegetation and animals change.
Mountain tops in the south are like northern
islands in a sea of plains or desert. On the
south side of a mountain each zone is higher.*

springtime flowers enrich the landscape; and various cacti. Animal life consists of lizards, snakes, desert rodents, and numerous birds. The Upper Sonoran takes in the semi-arid regions at the lower elevations of the mountains, where the vegetation consists of desert grasses, sagebrush on the drier slopes, and willows along the rare watercourses. Coyote and elk are at home in this zone.

The Foothill Zone is dominated by the "pygmy" forest, consisting of piñon pines and junipers, with wide areas of such shrubs as mountain mahogany, scrub oak, serviceberry, and chokeberry. It supports a large rodent population of rats, mice, chipmunks, and squirrels, and various reptiles. In the Montane Zone, the lowest region of true forest, the botanical indicator is the ponderosa pine. These pines grow in open, parklike formations, sometimes in the company of Douglas fir, and are replaced by scattered stands of lodgepole pine and aspen in areas first cleared by fire or by other natural or man-made disasters. A rich population of large animals and spectacular birds lives in these forests. Nearer the top of the mountain, in the Subalpine Zone, the Douglas fir begins to disappear and the thick, deep-green, majestic forest of Engelmann spruce and subalpine fir reach up to the timberline. These forests are also sometimes spotted with intrusions of lodgepole pine and aspen. This is the range of the moose. Some of the upper reaches of the Rockies are high enough to have a permanently frozen subsoil. This is the tundra, with lichens, sedges, grasses, and magnificent miniature flowers in extensive, rock-strewn meadows. The mountain goat is thoroughly at ease on the barren slopes above the timberline.

Few species, however, are strictly confined to one life zone. Animals move freely up and down the mountainside. Many plants are distributed over extensive regions. But a fairly large number are confined to a limited area that loosely coincides with a particular life zone. Zonal identification can generally be made quite accurately from the surrounding vegetation. In the mountains, innumerable combinations of environmental factors have a profound effect on the evolutionary process. Here plants and animals must become adapted to conditions that may range from desert to tundra. As life moves from the warm lowlands toward the Arctic climate of the mountaintops, plants and animals must evolve in harmony with the more and more demanding requirements of the higher and higher altitudes. At the end of a long period of trial and error, one group evolves with the combination of qualities best suited for survival in a particular region.

Yarrow, an aromatic plant with flat-topped white flower heads, grows from the lowest valleys to well above the timberline. On the way up from the lowlands to the highlands, from a nine-month to a two-month growing season, this one- to three-foot-high plant with one-and-a-half-inch leaves dwindles to only three to ten inches tall with tiny leaves. The open, shrubby forests of piñon and juniper of the dry lowlands give way to the ponderosa pine in the next higher zone, where the moisture is a little more abundant. But in the still higher areas, the pines fade out, and spruces grow, because their shallow root systems are compatible with the greater precipitation and the thinner soil of the high altitudes.

Separate species of gallinaceous birds (the order that includes the common domestic fowl, as well as pheasants, grouse, and turkeys) have colonized different zones because each has evolved with characteristics best suited to that environment. In the evolutionary process of natural selection, slight genetic differences among individuals tend to be perpetuated if those differences are advantageous in a particular environment. These differences may become greater and greater, and if one group is kept apart from the other, then over a million years or so, members of one group may not be able to interbreed with members of the other group. At this point a new species has evolved. As a result of the innumerable permutations of environmental factors, a climb from lowland to mountaintop encompasses a remarkable number of ecosystems, some simple and others very complex.

TALLGRASS PRAIRIE

A few centuries ago, the American prairies extended over the central plains of the continent as a vast expanse of grass, from the Rockies to the eastern forests and from Canada to the Gulf of Mexico. The natural origins of these plains go back to the time when the Rocky Mountains, lifted from the earth by enormous volcanic forces, began to capture the moisture in the air masses that drifted from the Pacific Ocean. With the creation of the new Continental Divide, the rivers and streams that flowed to the east began the long process of distributing quantities of sediment over the open plains. All this debris was first deposited in fan-shaped accumulations that ultimately merged into a single, gentle plain as the meandering rivers and streams changed their courses and drifted from side to side. In the northern regions, rich deposits of debris were added, layer by layer, as the glaciers retreated periodically with changes in the world's climate.

A prairie is generally defined as an extensive tract of flat or rolling land, broken up by threading watercourses and destitute of trees, with a deep fertile soil chiefly covered with grasses. The characteristics of the grasses that originally covered the prairies were determined in large part by the measure of rainfall. There were three basic grass zones: the eastern, the western, and the transition zone between them. The lowlands from Kansas eastward, where the rainfall is most abundant and the climate quite salubrious, were dominated by tallgrasses: big bluestem grass, switch grass, and Indian grass. In the high-plains region west of the one hundredth meridian, where the land rises toward the wall of the Rockies, the predominant species were shortgrasses: low western wheatgrass, buffalo grass, and blue grama. In the middle zone, the so-called mixed grasses—little bluestem, needle and thread, and prairie

June grass—were prevalent. Apart from rainfall, variations in temperature, soil composition, and topography contributed to the formation of a great many diversified habitats. In all, several hundred species of grasses, which overlapped and intertwined, covered more than a million square miles and bound the soil into a continuous sod. The prairies and their resources seemed endless.

Today it is estimated that about 50 percent of the short and midgrass prairies have survived, but that only 1 percent remains of the 400,000 square miles of the original tallgrass prairie. With the introduction of John Deere's steel-moldboard plow, the areas east of the one hundredth meridian succumbed to the "sodbusters." These regions now represent the paramount agricultural asset of the nation and a source of food for many other parts of the world as well. The rolling plains of Kansas and Nebraska are planted to winter wheat, seeded in the fall, semi-dormant in winter, maturing in the spring, and harvested in June or early July. In the north, where little vegetation can survive the long, bitter winters, summer wheat is sown in the spring and reaped in August or September. Broad, open cornfields cover major portions of Iowa and Missouri. There are rye, barley, oats, and other crops besides wheat and corn. Crop rotation, terracing, soil enrichment, and other techniques had to be learned over many years of hardship before the land would yield its present-day bumper crops.

Each year more prairie land is plowed and planted to marketable crops. Additional areas fall before road building and urbanization. The largest relatively undisturbed prairie in the country is in the Flint Hills section of Kansas, covering an area two hundred miles from north to south and about fifty miles wide. Because the area has a shallow, rocky soil, most of it has never been plowed. These prairies are privately owned and, under moderate grazing practices, have survived in reasonably good condition. There are small scattered parks, refuges, and recreational areas maintained by states, counties, and private organizations in Iowa, Kansas, and Nebraska in which tallgrass communities have been preserved. The Nature Conservancy owns over 36,000 acres of grassland in twelve midwestern states. Most tracts are predominantly tallgrass prairie. The largest single preserve is the Konza Prairie near Manhattan, Kansas; there, on more than 8,000 acres, plant and animal studies are conducted by Kansas State University and scientists from all over the world. Remnants of tallgrass vegetation in its natural state can be found sometimes along railroad rights of way, fencerows, and roadways, and in parts of old settlers' cemeteries. But even these minimal fragments are disappearing, with the practice of "lawn-mowing" of roadsides, the use of chemical sprays, and the elimation of fencerows.

Much effort has been expended to preserve these precious relics of the early American landscape. Movements to establish a tallgrass national park have popped up periodically during the last fifty years, each time squelched by farmers, cattle ranchers, and developers, all of whom have had other uses for the land. The present leaders in the drive for a tallgrass sanctuary are supporting legislation that would provide for a combined national park and national preserve. This designation would permit the continuation in a substantial part of the area of current farming, ranching, and oil and gas operations pending a long-term program of land acquisition. The projected site is in eastern Kansas. Tallgrass can grow back to full height in two or three years, but the restoration of the original animal life, with a full complement of buffalo, antelope, elk, coyotes, and others, will take many more years.

A tallgrass prairie in full growth is a spectacular sight—a sea of vegetation waving in the wind, with green, yellow, and brown ripples catching the bright summer sunlight. Big bluestem is the hallmark of the tallgrass prairie. In the low, moist prairies, it sometimes reaches a height of eight feet, with root systems that go six feet down into the rich soil. Switch grass requires less moisture, so it thrives on warmer and drier sites, and grows about one foot shorter than big bluestem.

Indian grass has the same basic rainfall requirements as big bluestem but is far less abundant than the other grasses. It grows mixed in with bluestem and almost as tall. Expansive settlers described the grasses as being tall enough to hide a man on horseback. Most of the grasses, however, rarely grow more than six feet tall, and the late-season grasses are generally much shorter.

Wildflowers have special problems competing against the dense carpets of grass. The struggle for dominance goes on primarily below the surface. Roughly only 15 percent of the total standing vegetation occurs above-ground, and 85 percent belowground. The various species can exist together only by sharing the soil nutrients at different levels and sharing the light at different heights. Once established, a wildflower can compete only if its fleshy root extends below the roots of the grasses. Some of the flower roots extend as deep as nine feet.

Beginning early in the spring and continuing into the last frosts of the fall, brilliant wildflowers add successive splashes of color. Before any other evidence of spring growth, the pale-lavender cup of the pasque-flower can be sighted. From then on into summer, sparkle is provided by purple clover, the pink inflorescences of prairie phlox, the long golden-yellow racemes of lupines, the terminal clusters of the violet flowers of spiderwort, the blue tubular flowers of penstemon, and the bright-orange flowers of the butterfly weed. In late summer and fall, tall species of wildflowers begin to rise to and above the level of the grasses. These include asters, sunflowers, goldenrod, coneflowers, and blazing stars. In the low moist areas are broad-leaved blue flag, water hemlock, smartweed, and swamp milkweed. In all, more than three hundred species of wildflowers add their sparkle to the green carpet of grass. To these must be added a few woody, colorful small shrubs, including the prairie rose.

For a long time the diverse flora of the grasslands sheltered a vast variety of animals. There were those that relied upon their speed in order to survive. These

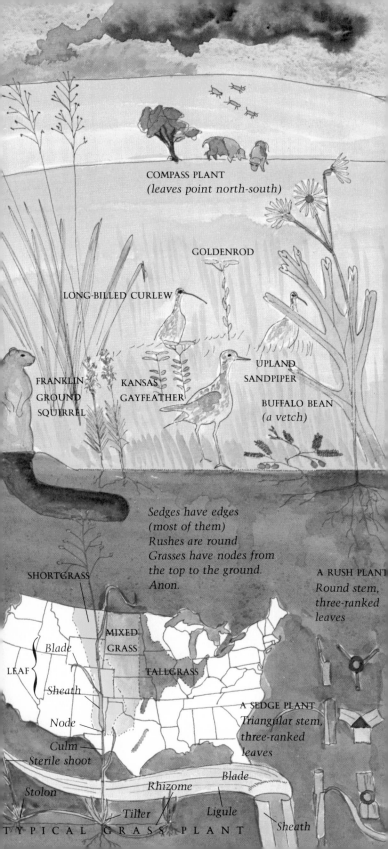

COMPASS PLANT
(leaves point north-south)

GOLDENROD

LONG-BILLED CURLEW

UPLAND
SANDPIPER

FRANKLIN
GROUND
SQUIRREL

KANSAS
GAYFEATHER

BUFFALO BEAN
(a vetch)

Sedges have edges
(most of them)
Rushes are round
Grasses have nodes from
the top to the ground.
Anon.

SHORTGRASS

A RUSH PLANT
Round stem,
three-ranked
leaves

MIXED
GRASS

Blade

LEAF

TALLGRASS

Sheath

Node

A SEDGE PLANT
Triangular stem,
three-ranked
leaves

Culm
Sterile shoot

Blade

Stolon

Rhizome

Ligule

Sheath

Tiller

TYPICAL GRASS PLANT

TALLGRASS PRAIRIE

BISON

SUNFLOWER

PRAIRIE CHICKEN

BIG
BLUESTEM

INDIAN
GRASS

SWITCH GRASS

A GRASS PLANT
Round stem, open sheath, two-ranked leaves

included elk, pronghorn, deer, and jackrabbits, as well as the coyotes and wolves that pursued them. Others, including mice, prairie dogs, and ground squirrels, hid in underground burrows and were hunted down or dug out by badgers, owls, hawks, snakes, and black-footed ferrets. The thick cover of grass provided ideal nesting sites for grouse, meadowlark, quail, and plover. There was an amply supply of grasshoppers, leafhoppers, grubs, beetles, and other invertebrates. The watery areas of lakes, marshes, and potholes supported populations of herons, grebes, cormorants, pelicans, and occasional families of whooping cranes. Of all the fauna, none was more majestic and more compatible with this environment than the enormous herds of buffalo, the American bison.

One of the threatened species of the tall- and mixed-grass regions that has attracted much public interest is the greater prairie chicken. Hunting and the destruction of its habitat have reduced this species to a small fraction of the original flocks. With some protection, they have shown an ability to adjust to conditions in fields and pastures that were formerly prairie country. Their annual springtime courtship ritual is an unforgettable prairie experience. In the pre-dawn silence, the birds gather on their dancing grounds, or leks, for their bizarre festival. The cock begins its display, strutting, inflating its orange neck sacs, and emitting "booming" sounds, all for the benefit of the hens on the sidelines. The prairie chicken is not a chicken but a pinnated grouse, and during its display the pinnae, the erectile black feathers on each side of its head, are thrust forward like a pair of horns. When another male arrives at the lek, the strutting and stomping turns into a clash of rivals. Their fight is sometimes ceremonial. But when the clash becomes real, the combatants tear at each other, collide in midair, and leave the ground littered with their feathers. The vanquished retires to other territory, and the watching hen approaches the victor. Each year, this strangely organized drama heralds springtime on the prairies.

Much of the rich wildlife community of the grass

country is gone. There are no open stamping grounds for bison, nor running room for the pronghorn. Pesticides have reduced the prairie falcon to a vanishing species. The wolf, coyote, and the prairie dog have been slaughtered and poisoned because of real or imaginary threats to crops and cattle. The long-billed curlew and marbled godwit are also threatened. Species after species have declined with the vast man-made changes in their habitats.

MIXED-GRASS PLAINS

West of the tallgrass prairie, the precipitation gradually decreases from about thirty inches to fifteen inches a year, and the wind and sun rob the land of much of its moisture. This is the mixed-grass plains, running like a ribbon through the central parts of North and South Dakota, Nebraska, Kansas, and Oklahoma and into the northern section of Texas. In these states great stretches of privately owned prairie may still be seen under various stages of grazing. In addition, numerous wildlife refuges, maintained by the Department of the Interior primarily for waterfowl protection, contain sections of mixed-grass plains with all the associated mammal, bird, and insect life. Other government property, purchased during the drought and Depression of the 1930s and later made part of the national forest system, is being restored to productive prairie vegetation. While much of this land is managed on a multiple-use basis, thousands of acres have recovered from past abuses and now support moderate grazing by cattle and the wildlife linked with grasslands.

On the southeastern edge of South Dakota's Black Hills, Wind Cave National Park preserves a section of the great plains—a prime example of a mixed-grass ecosystem, with a rich blending of grasses, a large assortment of spring and summer wildflowers, many typical prairie animals, abundant bird life, and outstanding vistas of unplowed prairie. Among the wildflowers are pasqueflower (South Dakata's state flower), scarlet globe-mallow, coneflower, mariposa lily, prickly poppy, and verbena. The park is a sanctuary for many species of mammals, the largest of which is the bison and one of the smallest the prairie dog. There are also pronghorn, elk, deer, coyotes, badgers, jackrabbits, and raccoons. The long list of birds includes meadowlarks, woodpeckers, warblers, chickadees, grouse, kingbirds, bluebirds, and magpies.

The grasses of the plains are not the majestic species that once covered the prairies to the east, or the shorter grasses that cover the adjoining steppe to the west. The mixture, a balance between short- and tallgrasses, includes a wide variety of species. Little bluestem, usually found in clumps, grows about two feet tall, with roots extending more than five feet below the surface. The white, hairlike tendrils on its underside retard evaporation. In especially dry periods, it reduces its exposed surfaces by rolling up the edges of its leaves. Needle-and-thread grass, known also as porcupine grass because of its sharp-pointed, barbed seeds, does most of its growing during the cool spring season. Prairie June grass prefers the cooler areas. These and others make a tough grass mixture, with a dense network of underground growth that once covered the plains with a matted sod structure that could withstand extended periods of drought.

For an animal to survive in the open prairie, it must be either strong or fast, or equipped to escape into an underground refuge. One of the most interesting of these burrowing animals, originally found mainly in the mixed- and shortgrass prairies, is the black-tailed prairie dog, an animal that has been the target of massive campaigns of poisoning and hunting for decades. Millions of their number have been exterminated because their hole-digging and grass-eating habits are considered incompatible with ranching. Those that survived are in isolated areas of their original habitat or in game reserves or national parks.

The prairie dog is actually a kind of ground squirrel; it owes its popular name to its sharp, barking call. It is a sociable animal that lives in large-scale communities called prairie dog towns. A typical town consists of various groups of prairie dogs, each group, or coterie, occupying and protecting its own small neighborhood within the town. Individual members of one neighborhood are not welcome in another. If a trespass is attempted, the intruder will be met by a resident standing on its hind legs, with its front paws raised over its head, and uttering a loud *yip* in defense of its territory. Members of each coterie cooperate with each other in constructing the burrow, and are often seen eating, playing, and grooming together. Yet prairie dogs communicate with their neighbors in different coteries, especially by their distinctive warning barks, whenever the town is threatened by a predator.

Prairie dogs are expert underground engineers, producing burrow systems at depths of ten or more feet that extend over wide areas. Different chambers are hollowed out for nesting, food storage, sanitary purposes, and as listening posts. The burrow generally connects two ground holes, each of which is surrounded by a mound constructed mainly from material brought up from the burrow and compacted into a firm structure. The mound is designed to protect the burrow from flooding. After the grass around the perimeter of the town has been removed, the mound becomes a perfect observation post from which to spot possible predators. The prairie dog's eyes are set high on its head. This adaptation allows it to see not only forward and to the side but also upward for possible predators. At the first sign of danger, a high-pitched call of alarm, different from the territorial bark and accompanied by a flick of the tail, is passed from mound to mound, sending the villagers scurrying for the safety of their burrows. If the predator follows down into the burrow, the interconnected ground holes provide an escape route for the prairie dog.

While the burrow is essential for survival of the prairie dog, each of its many predators has developed its own technique for overcoming the organization of the town. Winged predators such as hawks, eagles, and falcons rely on their speed to catch their prey in the open. Rattlesnakes and bullsnakes will follow a prairie dog down into its burrow. The badger claws his way into the ground to make its kill. The coyote has to rely on distractive maneuvers to succeed in its hunt. The black-footed ferret, which depends on the prairie dog for its primary food supply, follows the prairie dog down its burrow. It has become an extremely rare species with the scarcity of prairie dog towns.

Many different species of insects, birds, reptiles, and mammals make use of the mounds and burrows. The stability of the temperature and humidity provides an ideal climate for crickets, beetles, fleas, and spiders. The burrowing owl appreciates burrows that have been abandoned by the prairie dog or are infrequently visited by them. Field-mice and rabbits use the burrows for shelter and comfort. Ground squirrels feed on the insects that live in the burrows. Insects around the mounds attract meadowlarks and horned larks, which also use the mounds as singing perches. Snakes use the burrows to escape the heat of summer and to hibernate during the winter. Bison often visit the towns, destroying the mounds, loosening the surrounding soil with hooves and horns, and then wallowing in the loose earth to remove matted clumps of old fur and annoying insects. This wallowing destroys some of the grass, allows other vegetation to become established, and produces deep hollows in the earth that collect rain and become waterholes.

The largest mammal of the plains, which for centuries shared the same habitat with the prairie dog, is the bison, now extinct in the wild state. The original herds were surely one of the greatest animal congregations that ever existed. They were well equipped for life on the prairie. Winter blizzards, the great menace to domestic cattle, were no threat to the bison. They would face squarely into the storm and, moving their heads back and forth, would plow through the snow to find the grass underneath. Always on the move, they neither damaged nor overgrazed the prairie grasses. But their doom was inevitable once the westward-moving pioneers recognized the agricultural value of the tall-grass prairies and the cattle-grazing potential of the mixed- and shortgrass plains.

Today, bison herds are rare sights, preserved in Wind Cave National Park, Custer State Park, Theodore Roosevelt National Memorial Park, Yellowstone National

MIXED-GRASS PLAIN

BISON *wallowing in prairie dog town*

PRAIRIE DOGS

BURROWING OWLS *in abandoned* PRAIRIE DOG *burrow.*

BADGER

WESTERN WHEATGRASS

NEEDLE-AND-THREAD GRASS

PALE PURPLE CONEFLOWER

PASQUEFLOWER

PRAIRIE RATTLESNAKE

Immature

GOLDEN EAGLE

PRAIRIE FALCON

BLACK-FOOTED FERRET

ED-TAILED HAWK

LITTLE
BLUESTEM
GRASS

JUNE GRASS

SHELL-LEAF
PENSTEMON

PRAIRIE
CLOVER

GRASSHOPPER
MOUSE

Park, and in a few other public and private ranges. The present-day bison ranges are the only places to see grass prairies in something like their original beauty and richness. There these magnificent beasts roam the open country, the cows peacefully ministering to their hungry calves, and the bulls fuming, snorting, locking horns, and otherwise displaying their enormous strength, especially in the struggle for dominance during the rutting season. The present herds are pocket-sized remnants, bands of two hundred, compared to the sixty million bison that once covered the prairies. These herds now have to be carefully managed to prevent their outgrowing the limited available preserves.

If the bison is the strongest of the animals of the plains, the pronghorn (erroneously called "antelope") is surely the fastest. This animal is ideally adapted for life in flat, open grass country. With powerful legs, keen eyesight, and enormous endurance, it is a difficult quarry for even the most skillful predator. The pronghorn can be identified by its unique pronged horns, its tan-and-brown coat, the two white bars on its neck, and its conspicuous white rump patch. At the first sign of danger, the animal becomes tense, and the twin white discs on its buttocks flare out as a signal that is visible at great distances to the rest of the band. At the same time, a special gland throws off a musky warning odor. Others pick up the warning and repeat it. In a short time, the whole prairie is astir with these swift-moving animals in flight, their rump patches glistening in the sunlight. They rely on their speed to escape from predators, but they first congregate in a herd, apparently as an aid to survival, before stringing out in single file. Their principal enemies are coyotes, bobcats, and wolves; when cornered by one of these, the pronghorn will defend itself with its strong, sharp hooves. Hunting and the destruction of its habitat have brought the pronghorn to a precarious state. Small herds are being protected and managed in various public and private refuges.

The bison and the pronghorn are just two of the many wild creatures that have been unable to cope with the disturbing changes made by man in the primi-tive prairies. The impediments to survival are unyield-ing. For many, there is just no place to live. A few escape to the remaining remnants of their ancient hab-itats. But most, unable to adjust to altered conditions, simply fade away, leaving to the more adaptable species the struggle for survival in the depleted environments.

SHORTGRASS STEPPE

Moving westward from the Mississippi River, the land rises gradually and steadily toward the Rocky Mountains. At the edge of the central lowlands, in Kansas, the land has risen fifteen hundred feet, and it reaches a height of a mile at the eastern face of the Rockies. Rainfall decreases as the land rises to the west. The imaginary longitudinal line known as the one hundredth meridian is the approximate boundary between the rich eastern region and the arid western region of the plains. East of this line, the average rainfall is between twenty and twenty-five inches, most of it brought by warm air masses moving in from the Gulf of Mexico. The land west of this line is in the rain shadow of the Rockies; its average annual rainfall of only five to fifteen inches is brought by Pacific air masses that drop most of their moisture before they cross the mountains. Scanty rainfall, high winds, unstable climate, and periodic droughts are basic features of the steppe.

In the eleven Western and Pacific states (Washington, Oregon, California, Nevada, Idaho, Utah, Arizona, Montana, Wyoming, Colorado, and New Mexico) 622 million acres (or 83 percent of the land area) are grazing lands. No single activity or combination of activities has contributed more to the deterioration of plant and animal life than the nibbling mouths and pounding hooves of livestock. When the land was grazed by bison and other wild animals, little injury was done. The grazing was intermittent and widespread, so that the land had a chance to recover. But large herds of cattle confined in fenced areas can produce enormous damage. The basic evil is overgrazing. When it occurs in the spring, some of the grasses are trampled into mud, destroying the root systems that protect the land against erosion. In the drier areas, the trampling of the livestock compacts the soil. This results in a greater runoff of rainwater, an increase in the number and depth of gullies and ravines, and a lowering of the water table. In the end, the palatable perennial grasses and weeds with the highest nutritional value are replaced by scrub brush, less desirable annuals, and even poisonous exotic species. The vegetative cover is changed and with it the animal life that is dependent on it. The quality of the rangeland declines until it lapses into desert wasteland.

Of these millions of grazing acres in the West, half are privately owned; the other half is public land. Of the approximately 250 million acres of public land, about 170 million acres is rangeland. Much of this land was originally homesteaded but reverted to the government when it was abandoned in poor condition by hopelessly disillusioned farmers. Cattlemen moved in, overgrazed the land, and left little but dust in many areas.

Grazing regulations issued by the supervising government agencies were until recently administered to accommodate the ranchers. Legislation affecting the range lands was controlled by small subcommittees dominated by Western congressmen interested primarily in their own regional politics. In 1976, in an effort to reverse the progress of desertification, Congress passed the Federal Land Policy and Management Act, giving the Bureau of Land Management a multiple-use mandate with power to resist political and rancher pressures. This was supplemented in 1978 by the Public Rangelands Improvement Act, providing millions of dollars for "range improvements." Now new regulations have been issued, and management plans for individual ranchers are being formulated. Resistance from the cattle-raising community is inevitable.

Various remedial techniques for restoring marginal land have been introduced. These include hauling a chain between two tractors to remove sagebrush, plowing, reseeding, the heavy use of herbicides, controlled burning, and the fencing of substantial areas to give some pastures a rest. The hope is that the public lands will be returned to their original fertility, that the size of existing herds will not have to be reduced, and that some of the restored land will be dedicated to public purposes other than grazing. All this will depend on the adequacy of the appropriated funds and the steadfastness of the administrators in pursuing the legislative objectives.

The shortgrasses are particularly versatile in accommodating themselves to the arid conditions of this section of the plains. The dominant plants are the buffalo grass, low western wheatgrass, and blue grama. These have some special adaptations for survival and, in addition to their seeds, some interesting techniques of propagation that are used by other plants as well. In the tallgrass zone, roots of plants generally penetrate deep down to reach the permanently moist soil. Apart from its deep roots, buffalo grass develops a wide network of fine roots just below the surface. These can quickly absorb large quantities of water from brief spring showers. In the dry summer, the plant remains dormant. It also reproduces itself by means of stolons, thin stems that grow along the surface of the ground. Each stolon produces buds at fairly regular intervals. With good climate, these buds develop into new plants and guarantee a dense growth of sod. Western wheatgrass uses another technique for producing sod. It also sends out horizontal stems with buds. But these stems, called rhizomes, grow below the surface. The buds survive the drought that kills the parent plant, and grow into new plants as soon as beneficial weather returns. The extensive root system of blue grama, one of the so-called bunchgrasses, helps it survive the drought. Its method of extending its growth is by producing a clump of side branches from the base of the plant, known as tillers. The resulting "bunch" efficiently crowds out other vegetation and contributes to a compact cover of sod.

The ability of the grasses to reproduce by seeding as well as by various methods of vegetative growth makes for severe competition among the species, each struggling for its share of sunlight, water, and nutrients. Some dominate the dry slopes, some the warmer climates, and others the poorer soils. Each fits into a place the others are unable to occupy. Together they make a tight ground cover. Other plants can survive only if they can make distinctive adjustments. Some flowers can produce their seeds before they can be overwhelmed by the grasses. Others manage to extend their stems above the level of the grasses. Many complete their cycle of growth while the grasses are dormant. During periods of severe dry weather, however, the shortgrass plains become pockmarked with desertlike communities. Cacti, tumbleweeds, yuccas, sagebrush, and similar drought-resistant plants compete successfully with the native grasses. Once established, these invaders are not easily dislodged.

The climate of the shortgrass country is one of extremes. In summer, the temperature can reach 120 degrees. During the winter, it may drop as low as 60 degrees below zero. Wind is a constant factor. It blows almost steadily, drying the soil and driving sand and tumbleweeds before it in blustering whirlwinds. In the flat, open habitat where the grass is no more than a few inches tall, survival for animals depends, just as it does on the mixed-grass plains, on strength, speed, or availability of an underground refuge. Bison, pronghorn, and prairie dogs are at home here.

One of the animals of the shortgrass zone combines several survival techniques. This is the swift fox, the smallest of the wild canines, whose tan-and-yellow coat blends easily into the colors of the prairie grasses during all seasons of the year. Apart from its size and color, its identifying markings are the black spots on each side of its muzzle, and its black-tipped tail. Among its survival characteristics are high speed, good eyesight, and keen hearing. While the swift fox will usually dig its own den, it can reconstruct for its personal use a den or burrow abandoned by another animal. Unlike the burrows of the prairie dog, the entrances to the den of the swift fox show no evidence of the excavated dirt. Fenced grazing lands are prime locations for these dens. In areas broken up for farming, dens are often dug along fencerows or roadsides, but in these areas the animals are easy prey for trappers and hunters. The swift fox is a nocturnal prowler for food, returning to its den with

SHORTGRASS STEPPE

COYOTE

McCOWN'S LONGSPUR

CHESTNUT-COLLARED LONGSPUR

PRAIRIE ASTER

RED MALLOW

BLUE GRAMA GRASS

ORD'S KANGAROO RAT

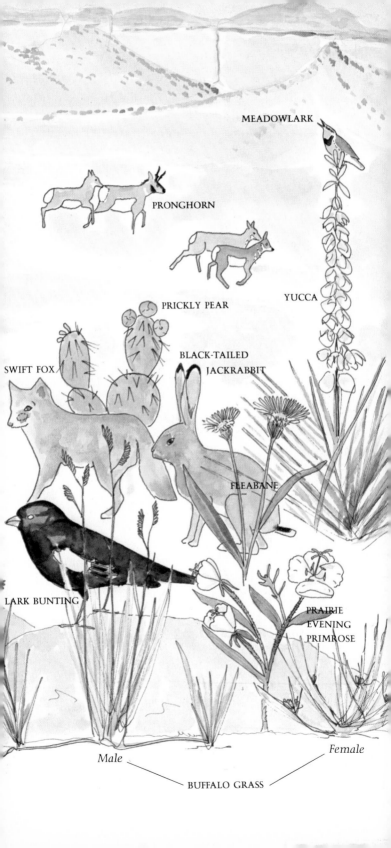

MEADOWLARK

PRONGHORN

YUCCA

PRICKLY PEAR

BLACK-TAILED
JACKRABBIT

SWIFT FOX

FLEABANE

LARK BUNTING

PRAIRIE
EVENING
PRIMROSE

Male

Female

BUFFALO GRASS

the first rays of dawn and remaining there for the rest of the day. It feeds upon rabbits, rodents, birds, and insects, and the diversity of its diet is an important factor in the control of exploding populations of other small prairie animals. Its principal predators are eagles and coyotes.

Loose soil and abundant vegetation make the short-grass plains an ideal habitat for many of the rodent burrowers. Their endless burrowing is beneficial. It opens the soil, permitting the easier penetration of air and water. In large numbers, these small animals in turn become enormous consumers. All year round, both night and day, an abundant population of prairie voles scurry along surface runways and extensive underground burrows, subsisting on seeds, roots, and leaves. Their grayish to dark-brown bodies and yellowish bellies blend well with the sandy soil of the short-grass zone. Another widely distributed rodent is the deer mouse, large-eyed, white-footed, and generally lighter in color than the vole. It is primarily a nocturnal animal, skittering through the trails and tunnels, supplementing its vegetable diet with a variety of insects.

One of the most active excavators is the pocket gopher, hustling day and night throughout the year, extending its underground tunnels in various directions to get at new sources of food. Its presence is readily detected by the mounds formed with the earth pushed out during its digging operations. Its tools are two pairs of powerful exposed incisor teeth and strong forefeet with large, curved claws. The loosened earth is pushed along in the main tunnel with the palms of its forepaws and finally discharged through the opening of one of the side tunnels. The pocket gopher is a highly specialized animal, with a stout, compact body designed in all its aspects for efficient digging. The "pocket" part of its name refers to its fur-lined external cheek pouches, used solely for transporting food. Its underground home has galleries for food storage and separate quarters for sleeping.

Rabbits are also prodigious consumers. The enormous appetite of the black-tailed jackrabbit, the common jackrabbit of the grasslands and the open plains, makes substantial inroads into the food supply of live-stock. Most of the day is spent in its lair. In the evening and early morning it ventures out to feed, traveling con-tinuously and leisurely in short hops. It relies primarily on its huge, thin, almost transparent black-tipped ears to detect the presence of an enemy. When sitting still or moving about slowly, its ears are often fully erect and shifted around to catch the slightest sound. When pursued, the jackrabbit can travel rapidly, its ears flattened back against its shoulders. In the northern plains, the white-tailed jackrabbit is one of the active grass eaters. Its brownish-gray summer coat turns white in winter.

If not for predators, the rodents and rabbits, by their proliferation and voracious habits, would in time de-nude the grasslands. Protective coloration, nocturnal ac-tivity, burrowing and tunneling, speed, a keen sense of sight and smell, and other adaptations for survival are not enough protection for these small animals. Their predators sharpen their own senses and techniques of detecting the slightest change in coloration or the small-est movement of an animal. Owls and hawks are con-stant hunters of mice. Just before dark, they can strike an unwary pocket gopher the instant it leaves the en-trance to its tunnel. The badger is a great digger on his own. Badgers and snakes can pursue a burrowing ani-mal right into the ground. Snakes are plentiful, and they prey on all the small mammals. The principal foods of the coyote are rodents dug from their burrows, or rabbits run down in the open. When small animals are scarce, the coyote will turn to bigger game. When pickings are lean, the coyote can subsist between kills on vegetable matter, insects, and carrion. The adapt-ability of its diet, its large litters of pups, and its cun-ning ways account for its survival and the maintenance of its numbers despite man's unremitting campaign to exterminate it. As a class, the predators are more effi-cient controllers of rodents than traps or poisons.

COTTONWOOD BOTTOM

The literature of the early West makes many references to cottonwood trees. The watercourses and creek bottoms at low and medium elevations were often lined with cottonwoods. For travelers across the vast grass country, the sight of these trees marked the location of water and a place to rest. Los Alamos and the Alamo were both named for cottonwood groves. Settlers soon learned that the cottonwoods grew rapidly and that they could be propagated quickly. Where there was sufficient moisture in the open country, rows of cottonwoods were planted as windbreaks. As towns developed in the mountain foothills cottonwoods were used as ornamental trees in the streets and as shade trees in the parks. In canyon country, their seeds drift through the air and float down the streams, leaving behind a trail of saplings. They succeed wherever there is water, even if their roots must stretch deep and wide to reach the supply. Most of these trees are broad-leaved cottonwoods. Along streams higher up in the mountains, the narrow-leaved cottonwood may be seen at the edge of the coniferous forests. It is distinguishable from the broad-leaved species by its narrow crown and willow-like leaves.

There is no mistaking the broad-leaved cottonwood. The grayish-brown bark of a mature tree is thick, rough, and deeply furrowed. Its trunk, two to five feet wide, is often forked near the base and clear of branches for half its height of fifty to seventy-five feet. The thick, wide-spreading limbs and drooping branchlets form a very wide, open crown. Its leathery leaves are broadly triangular, short-pointed, coarsely toothed, with thin yellow midribs and long, flattened stems. This stem structure keeps the foliage trembling in the slightest breeze.

In the spring, the ripe fruit on the female trees burst

open and release tiny dark-brown seeds that float away on tufted cottony hairs. The air is filled with what looks like snow. In the towns, the residents are annoyed as the cottony seeds get into their hair and clothes, catch in the screens of houses, and leave deep drifts of litter on streets. By this time, the budding limbs have been transformed into pale-green, lacy structures. With the hot weather, the leaves darken, spread their wide circles of shade, and rustle with the wind. After midsummer, the massed foliage turns orange-yellow. In the canyons, they look like a fire racing along the watercourses. In winter, the trees become dormant, standing as unimpressive tangles of naked branches, often infested with a parasitic mistletoe, with the sunlight warming the earth beneath them.

Where the land becomes low and flat, a sluggish stream or a wandering creek may build up a pond around which a dense thicket of deciduous trees, shrubs, and herbaceous plants takes hold. Standing among the cottonwoods and belonging to the same family are the widely distributed willows, some tall, others small, and a few tiny creepers. One of these is the peachleaf willow, a tree that can achieve a height of sixty feet or more. Its leaves, light-green above and grayish below, are lance-shaped, with finely toothed margins that taper to a narrow point. In all species of willows, the leaves are alternate and much longer than they are wide. All willows can be recognized by the single cap, or scale, that covers each bud; buds of other trees are covered by two or more scales that separate when growth begins. A smaller tree that also prefers the moist soil around ponds or along streams is box elder. It has opposite pinnately compound leaves with three to seven leaflets that are coarsely toothed or shallowly lobed and long-pointed at the tip. It is a member of the maple family, and its characteristic winged seeds hang on the branches in clusters throughout the winter. Other plants that prefer this environment but which may grow as trees or shrubs are hackberry, chokecherry, water birch, and mountain alder. The sweet fruit of the hackberry is a wildlife favorite and was consumed by the Native Americans. The fruit of the chokecherry, as suggested by its name, is not very palatable. The water birch has dark-brown, nonpeeling bark, and twigs covered with resinous knobs. The cones, or strobiles, on all birches disintegrate when mature, scattering their scales and tiny seeds. In the fall, the leaves of this birch turn yellow, contributing their beauty to the surrounding landscape. The leaves of the thin-leaved alder, a plant associated with the birch, have orange-brown midribs and rusty hairs on the reverse side. Unlike birch cones, the cones of the alders do not disintegrate.

In some places, these trees and shrubs are crowded out by the spreading branches of the tamarisk. This shrub or small tree, which looks somewhat like a juniper, was introduced to the West from Europe at the beginning of the century, both as an ornamental because of its attractive pinkish-white flowers and as an agent for erosion control because its deep roots hold dirt during heavy rains. The introduction of this exotic has lamentably proved to be too successful, for it is a ragged and unkempt plant when not in blossom, and a fast-growing and tough competitor that has adapted to the environment, spread along the watercourses, and become a formidable threat to the native species.

These freshwater areas are among the richest of plant and wildlife habitats. Storms wash mineral-rich soil and life-giving debris down from the uplands. Plant and animal matter settles and rots at the bottom of the pond. Around the borders of the pond, marshy areas begin to emerge. The soil and water are enriched by microscopic organisms and by many other aquatic plants. Some float freely, some are submerged, and others, rooted in the mud, emerge above the water. Cattails, sedges, rushes, and grasses settle along the edges of the pond, where reptiles, frogs, salamanders, and turtles are active. There is plenty of food for worms, beetles, water striders, and backswimmers, and for the larvae and nymphs of various insects. Tadpoles, minnows, and fish are abundant. Bees, wasps, and butterflies seek out the shoreline vegetation. Birds are attracted by the water and by the insects, seeds, and various water-tolerant plants. Wildlife in nearby wooded areas comes down to the edge of the water where the entangled growth frustrates some of their predators. Every niche in this fertile environment is occupied, and survival for all species depends in some cases upon sharing common living spaces and food supplies.

The herons have perfected the technique of sharing. Different species nest in the same trees but at different levels and feed on the same basic food supply but at different depths in the water and at different times. The stately and slow-moving great blue heron often stands motionless with one leg raised, knee-deep in the water, as he watches for the telltale ripple of a fish, crustacean, or other morsel. The snowy egret, the bird that was almost hunted to extinction by the plume merchants of the nineteenth century, stalks its prey by stepping gingerly through the shallower waters. The "little" green heron, almost too short-legged to wade, does most of his fishing from the muddy pond margin or while perched on an overhanging poolside branch. The night herons roost in the trees during the day and do most of their hunting at night. The American bittern, with its speckled brown feathers, makes its home in the heavy cover among the grasses and sedges. When approached, it relies on its concealing coloration and freezes into position with its head and neck stretched in the air like a cattail stalk.

Other species of birds occupy other niches in this wet habitat. Wherever open water is surrounded by dense vegetation, the Virginia rail is at home. This colorful bird blends into the vegetation and, with its laterally compressed body, can slip easily between the cattails and rushes. Its nest is made of the stalks of these plants and is fastened to the vegetation above the level of the water. The two common blackbirds are well represented in the marshy areas. The redwing male has handsome military shoulder patches. The bright color of the yellow-headed male stands out sharply against its all-black body. While they establish their colonies in separate areas, the redwings preferring the drier stands

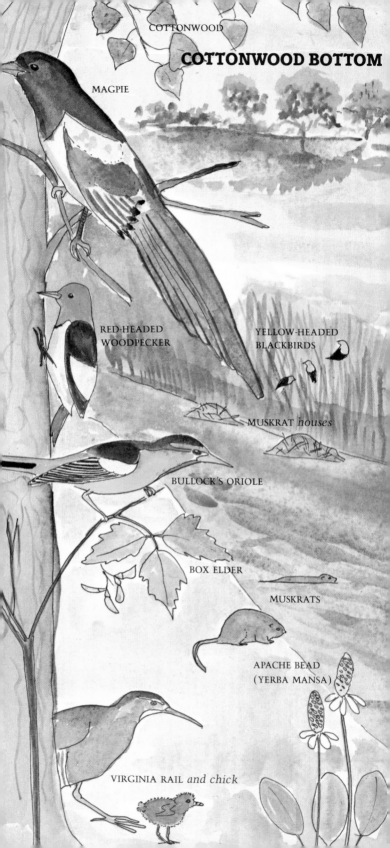

COTTONWOOD

COTTONWOOD BOTTOM

MAGPIE

RED-HEADED
WOODPECKER

YELLOW-HEADED
BLACKBIRDS

MUSKRAT *houses*

BULLOCK'S ORIOLE

BOX ELDER

MUSKRATS

APACHE BEAD
(YERBA MANSA)

VIRGINIA RAIL *and chick*

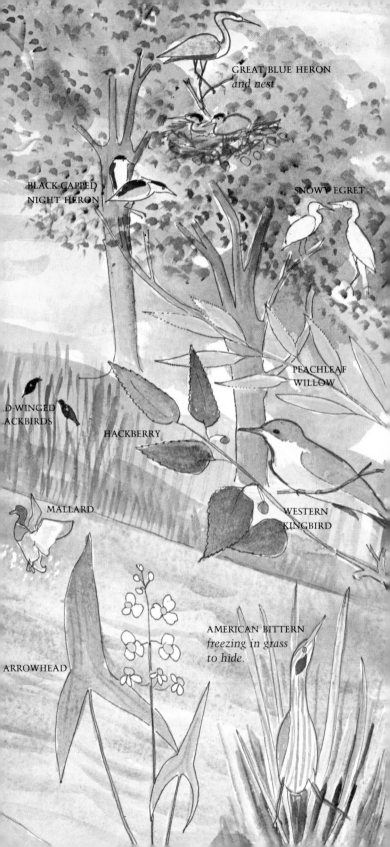

GREAT BLUE HERON
and nest

BLACK-CAPPED
NIGHT HERON

SNOWY EGRET

PEACHLEAF
WILLOW

.D-WINGED
.ACKBIRDS

HACKBERRY

WESTERN
KINGBIRD

MALLARD

AMERICAN BITTERN
*freezing in grass
to hide.*

ARROWHEAD

of cattails and the yellow-headed blackbirds the protection of the watery areas, both join together in a mob attack on any approaching predator. The sight of a flock of blackbirds rising to circle the marsh and then settling down to resume their feeding on insects and seeds is a thrilling early morning sight. The black-billed magpies require trees for nesting and are often seen flying among the cottonwoods. They are unmistakable in flight, the long black tail contrasting with the white wing patches and white belly. The orange-and-black Bullock's oriole will hang its basketlike nest on a branch that stretches over the pond. The western kingbird, pale gray above and yellow below, will place its nest in a tree or bush near the pond. If there are some dead trees among the cottonwoods, woodpeckers will use the cavities for nesting and for storing food. The open pond attracts a large number of ducks. Most rely on plant food, but some will also take animal matter such as insects, frogs' eggs, and mollusks. The mallard, the ancestor of the domestic duck, is one of the most widespread species. The male is recognized by its vivid green head, white neck-ring, and chestnut breast. The female has a mottled brown body and white tail.

Indians and early settlers made use of many of the plants around the edges of the ponds. The rootstocks and young shoots of the familiar cattail, identified by its brown cylinder of tightly packed flowers, were considered delicacies. The arrowhead, whose generic name comes from the arrow-shaped leaves of some species, has a tall stalk with small, three-petaled white flowers that usually occur in whorls of three. Its popular name, "duck potato," refers to the edible tubers that were dug out of the mud. Yerba mansa, or Apache bead, has a spike made up of tiny, sparkling-white flowers. Its aromatic rootstock was used as a remedy for a variety of ailments.

Muskrats find their way to the source of all this rich organic produce. Bulging above the shallow water are large mounds of debris, constructed of a network of cattails, reeds, grasses, and other matted vegetation. These muskrat lodges rest on piles of roots or mud and are about four feet high. Most are constructed with just one large nesting chamber with one or more underground entrances. Muskrats spend much of their time in the water, and their partially webbed hind feet, rich brown coats, and long, flattened tails are their adaptations for swimming. Their food consists mainly of the roots and leaves of aquatic plants, and of some fish, frogs, and crustaceans. Most of their food is towed out to eating platforms. These floating rafts, constructed of the same plants used in their lodges, are often littered with uneaten plant cuttings. Young muskrats are self-sufficient within about two weeks after birth and are then driven out of the lodge to care for themselves. Raccoons, foxes, coyotes, mink, and skunks are constant predators, often ripping open the lodges to get at the young. Droughts or flooding may also take their toll of the muskrat population. But there is a rich supply of the requisites for life, and a sufficient number of all species survive to play their part in the balanced community of the cottonwood bottom.

THE BLACK HILLS:
NORTHERN CONIFEROUS FOREST COMPLEX

The Black Hills are an isolated group of mountains in western South Dakota formed by the forerunner of the uplifting disturbances that produced the Rocky Mountains. The structure of this substrate is very similar to that of many parts of the Rocky Mountain chain. To the east are rolling farmlands and wheat fields; to the west are ranchlands that support millions of cattle and great flocks of sheep. The geological formation is a domelike, elliptical rock island covering nearly 6,000 square miles, extending about 120 miles north and south and about 40 to 50 miles east and west. The Black Hills are not really hills but mountains that rise several thousand feet above the surrounding high plains. The elevation of Harney Peak, a picturesque mass of pink granite, is 7,242 feet, the highest point east of the Rockies. Nor are the Black Hills black. Their color is the deep, cool green of the pines and spruce that cover them. They do have a dark appearance in contrast to the arid landscape that surrounds them, so that they seemed black to the Native Americans and settlers who approached them from a distance.

The Sioux tribes who moved into the plains early in the nineteenth century did not build any of their villages in the Black Hills. They hunted there because of the abundance of wildlife, but believing the Black Hills were inhabited by powerful gods, they otherwise avoided the area except as a sacred refuge in which to conduct religious or ceremonial rites. Pioneers began moving into the region shortly after the Civil War. The discovery of gold in the Black Hills in 1874 by the Custer expedition touched off the great rush. By 1876, the Hills were overrun by white settlers, precipitating the last of the major Indian wars.

The whole area abounds with legends of the Old West and includes such attractions as Mount Rushmore

National Memorial with Gutzon Borglum's sixty-foot-high, massive tribute to four American presidents, Wind Cave National Park, Custer State Park, Jewel Cave National Monument, Spearfish Canyon, Deadwood Gulch, Crazy Horse Monument, and, to the east, the spectacular Badlands National Monument.

Situated near the geographical center of the United States and surrounded on all sides by high plains, the mountains of the Black Hills form an isolated and unique crossroads for the flora and fauna of the east, west, north, and south. Here the living organisms of the surrounding major ecological systems meet to create diverse and overlapping biological patterns. From the east has come a group of deciduous forest trees, such as bur oak, green ash, and elm, which occur as scrubby growth at lower levels along drainage areas. The west has contributed ponderosa-pine forests, with stands of lodgepole pine and aspen, which dominate major sections of the woodland areas. In addition to these, the second most prevalent conifer, the white spruce, comes to the Black Hills from the north, as does the paper birch. From the arid southwestern plateaus have come yucca and cactus and the cottonwood of the low streamsides. Surrounding these three associated forest complexes are the vast grasslands of the Great Plains. However, the rainfall is higher in this region than on the arid plains below. An aerial view of these mixed communities would reveal a series of fingerlike extensions of pine forest, intermixed with deciduous elements along drainages, that reach from the dense growth of the upper elevations down into the valleys and on to the foothills that border on the high, open prairies.

The white-spruce forests generally occupy the north slopes in the higher, moister regions. Their presence in the Black Hills is especially interesting because they represent a solitary island of a species normally found several hundred miles to the north. The one-inch needles of the white spruce are rigid but not prickly, and are generally crowded on the upper side of the branch. Because of the smell of the crushed needles, the white spruce is often called "skunk spruce." The cones of all species of spruce are pendant. Those of the white spruce are from one to two-and-a-half inches long. The scales are thin and flexible with smooth, rounded margins. The outer bark of the tree is ash brown; the inner bark is silvery when first exposed. The lower branches of the tree are usually covered with a silvery lichen called old man's beard, which looks like tiny clumps of shrubbery and is often mistaken for Spanish moss. One of the deciduous trees in the mixed-forest areas of the northern slopes is the paper birch, the tree usually used by Indians for making their legendary canoes. This birch does best in the cool, moist, shaded areas of the forest. The bark on the trunks of mature trees is white, marked with dark blotches, and peels in long horizontal strips that curl up at the ends. The oval-shaped leaves are two to three inches long, with a rounded base, pointed tip, and double-toothed margins.

The undergrowth of a spruce forest is typical of a northern coniferous forest. In addition to many flowering plant species, it includes ferns, mosses, lichens, and grasses. A common flowering plant is the pipsissewa; its stiff, glossy, dark-green leaves grow in whorls on a stem supporting several open pink flowers or buds. Among the many other flowering plants are a variety of orchids, including the uncommon Venus'-slipper, or fairy slipper, and the yellow lady's-slipper. The Venus'-slipper is the only pink single-flowered orchid in the region. Its showy, drooping flower appears soon after the snow melts and can be found late into June. Both slippers derive their fanciful names from the inflated lip characteristic of various species in this group.

Specimens of the spectacular big-game animals can be seen in nearby parks and preserves, either singly or in fairly large herds. These are the areas in which bison, pronghorn, wapiti, and mountain sheep have been saved from extinction. In the Black Hills themselves, white-tailed deer are common sights, while mule deer, wapiti, and pronghorns are to be seen in the surrounding high plains. One of the rarer animals is the Canada lynx, a species ranging much farther north, which was probably trapped in the Black Hills by retreating ice sheets. Among the smaller mammals, the northern flying squirrel is a resident of the white-spruce forests. A folded layer of loose skin on each side of its body between the front and hind legs permits this nocturnal squirrel to glide from tree to tree. Another nocturnal animal of the spruce forest is the marten. It makes its den in a hollow tree and lives on a varied diet of smaller mammals, insects, birds, fruits, and nuts. There are many other species of mammals. Together they make a complex mixture reflecting the various habitats that overlap in the Black Hills.

As in the case of the plant life, bird species from widespread geographic environments meet and overlap in the Black Hills. Among the permanent residents are hawks, eagles, owls, woodpeckers, jays, nuthatches, magpies, grouse, juncos, and sparrows. The summer residents include vultures, killdeer, sandpipers, nighthawks, swallows, wrens, bluebirds, vireos, warblers, buntings, and western tanagers. Many of the transient species prefer the moist and watery habitats. These include grebes, Canada geese, shovelers, scaup, coots, and widgeons. Swainson's thrush, a summer resident, and the evening grosbeak, a winter resident and spring and fall transient, prefer the mixed coniferous-deciduous habitats, while the ruby-crowned kinglet, a summer resident, chooses the conifers. Because of this diversity of species, the region provides superb bird-watching experiences.

The Black Hills offer a number of choice locations for those interested in rocks, minerals, gemstones, and fossils. There are many historic mines in the area, including the nation's largest gold mine, the Homestake, in Lead, South Dakota. Apart from gold, these mines have produced copper, silver, lead, tin, feldspar, mica, and quartz. Rose quartz, the attractive pink gem mineral, is found in large deposits. There are also large deposits of iron, gypsum, alabaster, and limestone. Rock collecting is allowed in some areas administered by federal authority. Fossils and agates are found in the sedimentary limestone. The nation's greatest storehouse of vertebrate fossils, in nearby Badlands National

THE BLACK HILLS:
NORTHERN CONIFEROUS FOREST COMPLEX
An isolated island of a species found several hundred miles to the north. High elevation and north-facing slopes.

EVENING GROSBEAK

VENUS'-SLIPPER, or
CALYPSO ORCHID

CANADA LYNX

PIPSISSEWA

NORTHERN FLYING SQUIRREL

WHITE SPRUCE, *also called "skunk spruce" because of the smell of its crushed needles.*

PINE MARTEN

PAPER BIRCH

RUBY-CROWNED KINGLET

SWAINSON'S THRUSH

YELLOW LADY'S-SLIPPER *(An orchid)*

Monument, contains the remains of dinosaurs, saber-toothed tigers, camels, and three-toed horses. Small invertebrate fossils of ancient water and marsh creatures and many plant fossils are widely distributed. No collecting is allowed in the Badlands National Monument.

There are a large number of spectacular caves in the Black Hills, resulting principally from the work of rainwater in the thick limestone formations that underlie the surface rocks. Limestone is a fairly soluble rock. As rainwater descends it picks up carbon dioxide from the atmosphere to form a weak carbonic acid that seeps through the joints and planes in the bedrock and dissolves the limestone. Good subsurface drainage is an essential prerequisite for this process, and this is ideally provided by the domal structure of the Black Hills. The existence of chambers and alleys in caves indicates that some of the excavation work is also done by underground water moving in all directions as new water filters down from the surface. The cave is complete when the water level drops low enough to leave dry tunnels and caverns. Most limestone caves are spectacularly decorated with molded pedestals, glittering flowers, bizarre draperies, ornate columns, drooping stalactites, and upthrust stalagmites. Wind Cave, in the national park on the southeastern flank of the Black Hills, named for the gusty currents of air that blow in and out of the cave mouth, contains a variety of unique crystalline formations. These delicate structures, arranged in the form of honeycombs, are called "boxwork." Besides these formations, there are unusual crystalline displays of "frostwork" and "popcorn." Miles and miles of the passages have been explored, but they have been preserved in their natural state, and those passages that are open to the public make the descent into this part of the earth an intriguing and awesome experience.

THE PYGMY FOREST

The belt of land lying immediately below the forest areas of the mountains is the Foothill, or Transition, Zone. At lower elevations, fingers of grassland or remnants of cactus flats reach up along the mountain flanks, but these soon give way to a shrub-type vegetation that dominates the landscape. The composition of this shrubby growth varies considerably in different parts of the Rockies. In each area, one type of community will yield to another at a higher elevation; local variation in moisture, wind, and sunlight will affect the prevailing species and the adaptive forms they assume. The Palmer Lake Divide in central Colorado seems to act as an east-west dividing line, with visibly different weather conditions and plant species to the north and south of this line. Different types of shrub formations are also to be seen in the valleys and in the more northerly regions of the Rockies.

On the lower mountain slopes in southern Utah and parts of Colorado and in northern Arizona and New Mexico, the most common shrub community is the so-called pygmy-conifer forest, in which the indicator species are piñon pines (nut pines) and junipers. The pines grow on these dry, rocky slopes in open formations and generally reach a height of no more than twenty-five feet. Intermingled with these pines are low-growing junipers. Vast areas of the southwest at elevations from five to seven thousand feet are covered with the contorted forms of these pygmy forests. Much of the land is semi-arid, situated on rocky plateau country that is pelted with violent summer rains. The weather is generally temperate, but winter at these altitudes can be cold with considerable snow, and summers may at times become intensely hot.

Of the different species of piñons, the most common in the four-state area is *Pinus edulis*, whose needles are

THE PYGMY FOREST

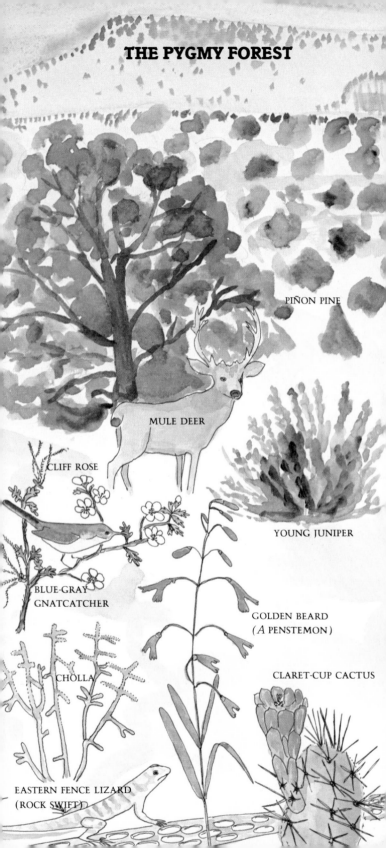

PIÑON PINE

MULE DEER

YOUNG JUNIPER

CLIFF ROSE

BLUE-GRAY
GNATCATCHER

GOLDEN BEARD
(*A* PENSTEMON)

CHOLLA

CLARET-CUP CACTUS

EASTERN FENCE LIZARD
(ROCK SWIFT)

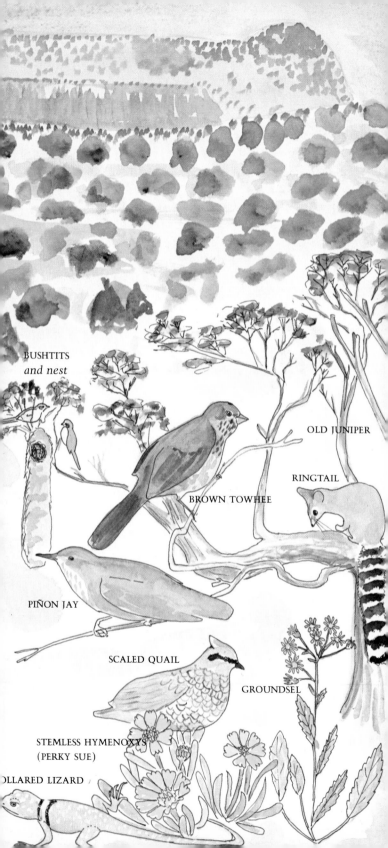

BUSHTITS
and nest

OLD JUNIPER

RINGTAIL

BROWN TOWHEE

PIÑON JAY

SCALED QUAIL

GROUNDSEL

STEMLESS HYMENOXYS
(PERKY SUE)

OLLARED LIZARD

borne in pairs. The wood of these trees was used by the Navajo and Pueblo Indians in the construction of their dwellings and for furniture and fenceposts, and as an important source of firewood. Piñon pitch was used as glue and for waterproofing straw baskets, as well as a dressing for wounds. But these trees were most important as a source of food. The scales of the cones open up every other year in late summer and release small edible seeds. These nuts were a basic part of the Indian diet and are today still harvested for family use. They are also harvested commercially and are sold in southwestern markets and elsewhere in gourmet shops as "Indian nuts" or "pine nuts." The harvesting is done by spreading blankets around the base of the tree and rocking the branches with a long stick to shake the nuts loose from the cones. The shell of the nut is soft and can easily be cracked with one's teeth. The nutmeat is sweet and has high nutritional value.

The junipers that grow among the piñons are short and gnarled; they are either one-seed junipers or Utah junipers. It is difficult to tell them apart, though the foliage of the one-seed juniper is gray-green and that of the Utah juniper yellow-green. The leaves of the young trees are very sharp-pointed and awl-shaped. In the older trees, the leaves are short and scalelike. The berries of both these species are usually one-seeded, while the bright-blue berries of the Rocky Mountain juniper usually have more than one. The junipers are also trees with many uses. The stringy bark makes excellent tinder for lighting fires and can be woven into comfortable sleeping mats. The wood of the tree can also be used for furniture and fenceposts. Although somewhat puckery, the berries can be eaten in an emergency and are used by Indians to flavor their food. These bluish-gray berries are actually modified cones in which a few fleshy scales enclose the juniper seed.

The plant cover is generally sparse in this rigorous environment. Most of the shrubs are ragged in appearance and spread low over the dry, rocky slopes. These include the cliff rose, which has small, dissected leaves, solitary cream-white flowers at the end of its short branches, and a long-tailed fruit; Apache plume has a white flower with leaves and fruits similar to those of the cliff rose; and antelope brush or bitterbrush has toothed leaves, yellow flowers, and a large brown seed. All of these shrubs are browsed by the game animals that come down from the higher regions of the mountains. Wildflowers cover the moister locations. An early bloomer is lamb's-tongue groundsel. It has small yellow flowers congested at the end of a single stem. There are clusters of large leaves at the base of the plant, with smaller leaves above. The cobweblike hairs fall off as the plant matures. During the summer, the scarlet flowers of the goldenbeard penstemon and the yellow flower heads of stemless hymenoxys can be seen on the open hillsides. Around the seeps and springs there are cottonwoods and other water-loving plants.

The pygmy-forest floor is littered with the chewed shells of piñon nuts and the rotting remnants of juniper berries, all evidence of animal activity. There is an ample supply of food for the abundant population of rodents, and the same predators that flourish in the slightly higher chaparral regions feed on the same prey in the pygmy forest. One of these predators is the elusive ringtail, a long-tailed raccoonlike cat that remains in hiding during the day and emerges only at night to hunt for small rodents and insects and to forage for berries and fruits.

There is plenty of food and shelter for the birds as well. A conspicuous year-round inhabitant is the piñon jay, which nests in both the pines and the junipers. These short-tailed, crowlike birds, crestless and steel-blue in color, congregate in raucous, noisy flocks and move about on the ground and in the trees searching for nuts, berries, and insects. Among the other birds in the shrubby forest are two that rely on insects for their food supply. The blue-gray gnatcatcher, a small, long-tailed bird with a white eye-ring, constantly flits about in the treetops in summer feeding exclusively on insects. The common bushtit, with its very long tail and very short bill, prefers the more arid brushy-type vegetation all year round and apparently obtains most of its moisture from insect food. The scaled quail is a common year-round bird of the drier areas. Its population fluctuates from year to year. It is gray, modestly plumaged, with a conspicuous white crest that accounts for its "cottontops" nickname. The brown towhee is a ground feeder, a permanent resident in the shrubby growth of creeks and streams.

In the morning and evening, mule deer, though extremely wary, will venture down to these watering areas. Apart from its large ears, the mule deer is distinguishable from the white-tailed deer by the black tip on its tail. It feeds primarily on twigs and shrubs, and, in the open parkland areas of ponderosa-pine forests, will graze on herbs and grasses. With many other large animals, it migrates up the mountainsides in the spring and down to the lower slopes in the fall. The attempt to protect a herd of mule deer early in this century is often cited to illustrate the possibly disastrous consequences of upsetting the existing balance of any ecosystem achieved by natural controls. Predators of the deer, including mountain lions, bobcats, and coyotes, were destroyed by game wardens and hunters. As a result of this overprotection, the mule deer population exploded. The available food supply was tragically insufficient for the swollen herd. Everything in sight and within reach was being devoured, the forest was being devastated, and thousands of deer were starving to death. When programs of control, including hunting in season, reduced the herd and maintained it within the capacity of the food supply, the forest began to recapture its former bloom and beauty.

The climate of the pygmy forest is still warm enough for reptiles, and its rocky areas provide convenient homes for them. One of these is the collared lizard, named for the band around its neck created by dark stripes. It is at home among the rubble, where it can find a ready refuge from midday heat and from intruders. It has a long tail, thin neck, and plump body, and its head is brightly colored with tinges of orange and yellow. It is an agile predator, opening and shutting its red-lined mouth and displaying its formidable teeth as it lunges after insects with a quick snap of its tongue. It feeds on smaller lizards as well. One of these smaller lizards is the eastern fence lizard, or rock swift. It is a good climber and will often be found in trees. As suggested by its name, it is hard to catch. Lizards are active during the cooler parts of the day and go down for the night into rock crevices or similar hideouts.

Much of the pygmy-forest country centers around the "Four Corners," the point at which the borders of Utah, Colorado, Arizona, and New Mexico converge on the vast tableland of the Colorado Plateau. This is a land of spectacular and unique scenery, a land of mesas and canyons filled with natural spires, needles, arches, bridges, and pinnacles. These, along with pygmy forests, can be seen in Mesa Verde National Park, Grand Canyon National Park, Canyonlands National Park, and Canyon de Chelly National Monument. This also is the home of the southwestern Indians—the Navajo, the Pueblo, the Hopi, and the Ute; relics of their ancient cultures are readily accessible. It is a land of ancient villages, pit houses, pueblos, and cliff dwellings, and of the early basket makers, potters, weavers, and silversmiths. It is a place of flaming color and magical variety, where daytime skies are filled in the summer with brilliant white clouds and where the nighttime skies support a canopy of sparkling stars.

THE CHAPARRAL

The pygmy forest of piñon pine and juniper is just one type of shrub vegetation found in the Foothill Zone. At slightly higher elevations, where the piñon-juniper is not fully developed, the chaparral community often occupies the mountainsides. Gambel oak is the most common indicator species, interspersed with a dense cover of low, woody vegetation composed mostly of tightly matted shrubs. *Chaparral* comes from the Spanish word *chaparro*, "evergreen oak"; the word *chaps* still describes the leather coverings worn as protection by those who ride through these thickets.

Prolonged periods of drought are a constant peril in these brushy areas. Many of the shrubs have various adaptive features that enable them to cope with this problem, including the evergreen habit: long taproots and leaves that are small, inrolled, glossy, and wax-coated, which prevents excess evaporation from the leaf surfaces. The plants may suffer during the hot spells, but most hold on until the next rainfall. Fire, however, is a more serious hazard, especially when the interlocking thickets have been dried out by the heat of a severe summer. Then they become an enormous tinderbox, ready to be reduced to charred and leafless splinters by a bolt of lightning or a careless man-made spark. But fire serves its purpose in the perpetuation of the shrubland. It clears out all of the debris and releases the minerals and other nutrients in the dead and living tissue. Many of the chaparral plants have adapted to this peril as well. Some have seeds that germinate rapidly when exposed to intense heat. Others, when burned to the ground, sprout new shoots from their old root system when the rains return. Soon the shrubs recover their former vigor, and in a matter of years, not decades, the chaparral has recaptured its domain.

The Gambel oak is not the majestic specimen of the

eastern forests. In the dry habitat of the Rockies, it has a stunted appearance. It never reaches a height of more than thirty feet and generally attains only half that size, with a diameter of no more than five inches. The Gambel oak is able to survive in its semi-arid environment because many trees sprout from a common root system that extends horizontally and can efficiently capture the limited rainfall of the lower mountainsides. But like the piñon pine, the oak supplies one of the most important sources of nourishment for the inhabitants of the wooded areas. Their acorns are rich in food value and are soft enough to be eaten by almost all animals. The Gambel oak is a member of the white-oak group, whose acorns are sweeter than those in the red-oak group.

There is a bountiful supply of berries and fruits, as well as acorns, in the chaparral. The berries and fruits are provided by a variety of plants. One of the most common plants in the whole Rocky Mountain region is the evergreen called kinnikinnik. This is a low, matted plant about six inches high, with brown trailing stems and small, pink, lantern-shaped flowers. Its leathery, dark green leaves and its bright red berries are important wildlife foods. In the drier areas of the foothill zone, compact clusters of dark red fruits of the smooth sumac add to the food supply. The leaves of this stout-stemmed shrub, which rarely grows above six feet in this zone, are dull green above and whitish on the underside. Orange and red berries are the fruit of another member of the sumac family—the three-leaved sumac, named for the three parts of its leaf. This shrub has several other regional names: it is called skunkbush because of the strong odor of its leaves, lemonadebush because its tart berries are used to make a lemon-tasting drink, and squawbush because split young branches were used in basketwork by Indian women.

Another berry-producing plant, a relative of the three-leaved sumac, is poison ivy. Its forms of growth vary. In the open areas, it is a ground-hugging vine. In the more sheltered areas, it appears as an upright shrub. In the forest it becomes a thick vine, using aerial rootlets to encircle and climb trees in its quest for sunlight.

ROCK SQUIRREL *digs nests beneath rocks but climbs to thirty feet and freezes for hours if in danger.*

SCRUB JAY in *"hiccup" posture warning of danger*

SCRUB JAY

GAMBEL OA
as t

BOBCAT

LONG-TAILED WEASEL

GRAY FOX

THREE-LEAFED SUMAC or "LEMONADEBUSH," "SKUNKBRUSH," "SQUAWBUSH"

Black tip of tail often attacked by predators, allowing weasel to escape

POISON IV

KINNIKINNIK

PACK RAT

CHAPARRAL OF OAK GLENS

MULE DEER

STRIPED SKUNK

BAND-TAILED PIGEON

WILD ROSE

SMOOTH SUMAC

MERRIAM TURKEY

COTTONTAIL RABBIT

GAMBEL OAK AS SCRUB

It can be recognized by its three leaflets, which are glossy when the plant is young. Its berries are white or grayish white, and its foliage, along with the other sumacs, turns brilliant shades of red, yellow, and orange in the fall. In no season, however, is it free from its poisonous oil.

In the moist areas, especially along streams, the prickly stems of the wild rose, or Wood's rose, support clusters of pink flowers in spring and early summer. The leaves and branches of these shrubs are browsed by animals, and the rose hips, or fruits, are an important part of the winter diet of several animals. The Indians used the hips as food, and they are still a favorite ingredient of homemade jelly. Like the wild rose, the entire plant of the serviceberry, or shadbush, is a palatable food for animals. Its dark-blue fruit was a staple in the diet of the Indians. Another widespread shrub in this foothill region is mountain mahogany, whose fruit, like that of Apache plume, has a feathery tail. It is not related to the tall mahogany trees of the tropics, and its name merely describes its hard and heavy wood, which makes an efficient fuel. Many of the scraggly shrubs of the pygmy forest appear in the chaparral as well.

Wild turkeys, slimmer than the barnyard variety, are fairly common in localized communities. The large rodent population of rats, mice, chipmunks, and squirrels are as at home in the chaparral as they are in the pygmy forest, and they all thrive on the carbohydrates and proteins in the available food supply. The most ambitious collector is the pack rat, or wood rat, known for the sumptuous supply of acorns and other food stored in its nest and for its habit of accumulating sticks, rubble, metallic objects, and other seemingly useless items as a barricade against invaders. The pack rat is a nocturnal animal, scurrying over well-beaten trails that extend considerable distances from its oversize nest. The big gray rock squirrel, with its long bushy tail, can climb trees, but builds its den beneath a boulder and is active on the ground most of the day. It too is an active storer of food. The nests of mice are everywhere—on the ground, in trees, and among rocks; many of these nests are also filled with seeds, acorns, and other nuts. The chaparral is also home for cottontail rabbits, who rely mainly on leafy green vegetation but will resort to buds, twigs, and bark in the wintertime.

All of these small animals provide in turn an equally rich food supply for their predators. The bulk of the gray fox's diet is rabbits and mice, occasionally supplemented with acorns, fruit, buds, and insects. The fox hunts mostly at night, stealthily prowling through the dense growth in search of its prey. It makes its den beneath boulders or in hollow logs but will often climb a tree to rest or to escape its enemies. It is distinguished by its pepper-and-salt coat and by the black stripe down the middle of its bushy tail. Another nighttime prowler, with a long neck and a long slender body, is the widely distributed long-tailed weasel. It kills its rodent prey by piercing the base of the animal's skull with its canines and hanging on until its victim expires. Weasel families often forage together, the mother followed by her young. The young of the striped skunk have similar nocturnal habits, trailing behind their mother on hunting expeditions for small rodents and insects. The presence of any skunk is usually first detected by its odor, but the two white lines in a V shape down the back of the striped skunk distinguish it from other species. The scent glands of the skunk are a good defense against molesters. Skunks, along with such mammals as raccoons, opossums, and badgers, do not hibernate in winter. Instead they retreat for long periods into a deep drowsiness. On warm winter nights, they will become active and leave their dens in search of food.

Among the larger nighttime hunters is the bobcat. Except for its short tail and tufted cheeks, it looks like an overgrown housecat. Most of the year, the bobcat feeds on small game such as rabbits and rodents; at times it will capture a fawn. In colder weather, it takes advantage of the movement of some of the larger animals down from the high coniferous forests to the food supply in the chaparral. Mule deer, feeding on acorns in the fall, are often ambushed by a bobcat. The phrase "lick his weight in wildcats" is a tribute to this ferocious fighter, who will not hesitate to do battle with enemies more than twice its size.

One common and most interesting bird among the scrub oaks is the scrub jay. It is distinguished from the more common blue jay by its lack of a crest, its blue rather than black collar, and the brownish patch on its back. It is a timid bird and will quickly disappear into the underbrush. It makes good use of acorns as food, along with insects and seeds, all found chiefly on the ground. Scrub jays are among those birds that have developed the social habits of cooperative breeding. In many cases the parent birds are assisted by their previous season's offspring, so-called helpers, in such essential activities as defending the nesting site, repelling predators, and feeding the young. Although the species is represented widely in the western states, in the East they appear only in the small scrub forest in southern Florida, more than two thousand miles away, where various species of stunted oak trees grow beneath taller sand pines. Another everyday bird in the oak groves of the chaparral, as well as in the pine trees of the pygmy forest, is the band-tailed pigeon, a large, swift-flying bird with a gray-tipped, fanned tail whose low-pitched call is similar to that of an owl.

Life in the chaparral favors the smaller animals. The stiff, interlaced twigs form a veritable jungle, ideal for snakes and lizards. Predators must learn to move skillfully through the thickets or go elsewhere. Large mammals, like deer, can come through the chaparral only on well-worn trails. It is no place for the backpacker. In this home of obstacles, the horse and chaps are requisites for travel.

PONDEROSA PARKLAND

Immediately above the Foothill Zone is the Montane, or Canadian Zone—the start of the true forest. As mentioned earlier, this zone begins at different elevations at different latitudes—at about 8,000 feet in the southern Rockies, 6,500 feet in the middle Rockies, and at only 4,000 feet in the northern Rockies, not far above the altitude of the high plains. Various types of forest communities have developed in this zone, depending upon latitude, east-west and north-south orientation, soil and rock conditions, rainfall and moisture, and stream and canyon locations. In many areas, different forest trees intermingle with one another, but for hundreds of miles on the southern sides of the mountains, there are dominating stands of ponderosa pine (western yellow pine) interspersed with groves of aspen and forests of lodgepole pine. Some of these trees intrude into upper or lower zones.

A ponderosa forest has a parklike quality. The trees are well spaced and of equal age, their strong trunks rise high into the sky, sometimes as much as 150 feet, and their broad, open crowns permit patches of sunlight to filter through to the ground. The trees thrive in dry, hot weather and on rocky soil, through which they send their long taproots. The trunks soar 50 feet, almost without tapering, before the first branch appears. The air is filled with the mild fragrance of vanilla, for the bark of the tree contains some of the same chemical used in the manufacture of artificial vanilla flavoring. On young trees, the bark is black or dark brown. On mature trees, the bright, reddish-brown, fire-resistant bark divides into irregular scaly platelets that seem to fit together like the pieces of a jigsaw puzzle. The needles are dark green, 5 to 10 inches long, generally in bundles of two that form tufts near the end of the branches. The oval cones, 3 to 6 inches long, have scales that are tipped with a stiff prickle. The undergrowth is sparse, for though the trees keep their distance, their expansive roots force out any competitive growth. They endure the hot, dry summers because their roots capture the spring meltwater, which sinks into the subsoil because it cannot run off the level forest floor. The ground is springy, littered with an accumulation of cones, needles, and the small platelets that scale from the bark of the trees. Openings in the forest are grassy meadows sparkling with displays of mountain wildflowers.

One of the animals firmly linked to the ponderosa pine is the tassel-eared squirrel, which builds its bulky nest in the trees. This is one of the largest and most colorful of the tree squirrels, with gray sides and reddish back; it is named for its long ears that end in blackish tufts, which are prominent except in late summer. It is confined almost exclusively to this type of forest, from which it gathers a year-long supply of food. In the spring and early summer, its principal food is the sweet, tender inner bark of the tips of ponderosa branches. The squirrel snips a few inches from the end of the young twigs, peels off the bark, eats the inside soft tissue, and then drops the woody core to the ground. In midsummer and early fall, when the new crop ripens, the squirrel is busy stripping the cones to get at the nutritious seeds between the scales.

A good example of a ponderosa-pine forest is near the north rim of the Grand Canyon National Park in the area known as the Kaibab Plateau. Here, the tassel-eared squirrel, called the Kaibab squirrel, has a black belly and a magnificent pure-white tail. Its relative in the ponderosa forest on the south side of the Grand Canyon, known as the Abert squirrel, has a white belly and a tail that is whitish only on the undersides. Many of the characteristics of each of these squirrels are the same, so that biologists are convinced that both species have a common ancestor. It is speculated that the Grand Canyon acted as an impassable barrier long enough for the evolutionary process to produce these clear differences in coloration. An animal restricted to a small area generation after generation will often become different from close relatives that move over extensive regions. It was first assumed that the white tail of the Kaibab squirrel had evolved to provide it with camouflage in the winter snows. But it does not appear to serve this purpose, so that its real adaptive value still baffles scientists. No explanation has yet been offered for the differences in the coloration of the bellies of these two squirrels. Since the forest on the Kaibab Plateau is surrounded on all other sides by desertlike areas, the Kaibab squirrel is not very mobile and is thus a phenomenon of a limited geographic region. Other species of tassel-eared squirrels do not live in isolated enclaves, and those in the other ponderosa-pine forests in Arizona, New Mexico, and Colorado are similar in general appearance.

There are many other animals in the ponderosa-pine forest. One of its inhabitants, the golden-mantled ground squirrel, is not solely dependent on the pine. This copper-headed squirrel with a white stripe on each side of its body has a varied diet of seeds, fruits, and insects. The mountain cottontail rabbit is strictly a plant eater and, along with squirrels and hares, supports a population of foxes, bobcats, and similar predators. Another plant eater, a characteristic animal of the more northerly forests, is the porcupine, a clumsy, heavy-bodied rodent. While primarily a nocturnal animal, the porcupine is an expert climber and may at times be seen during the day hunched up in the shape of a ball high in a tree. It feeds on buds and tender twigs, and its habit of stripping the bark off trees to get at the softer layers beneath results in obvious signs of its presence in the form of scarred tree trunks and a litter of broken branches. By its gnawing, clawing, and girdling, one porcupine can destroy a large number of trees in its lifetime. At birth, porcupines are armed with a full complement of soft quills that become hard and serviceable within minutes. These sharp, barb-tipped spines are formidable defensive weapons, and few pred-

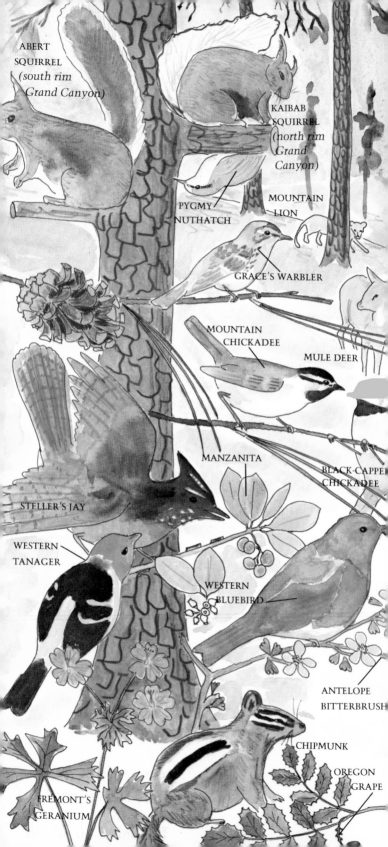

ABERT
SQUIRREL
*(south rim
Grand Canyon)*

KAIBAB
SQUIRREL
*(north rim
Grand
Canyon)*

PYGMY
NUTHATCH

MOUNTAIN
LION

GRACE'S WARBLER

MOUNTAIN
CHICKADEE

MULE DEER

MANZANITA

BLACK-CAPPED
CHICKADEE

STELLER'S JAY

WESTERN
TANAGER

WESTERN
BLUEBIRD

ANTELOPE
BITTERBRUSH

CHIPMUNK

OREGON
GRAPE

FREMONT'S
GERANIUM

PONDEROSA PARKLAND
6,500- 8,000 ELEVATION

PONDEROSA-PINE (YELLOW PINE)
Bark smells like vanilla

WALLFLOWER

PORCUPINE

GOLDEN
BANNER

MONKSHOOD

riodic
all fires
rn
dergrowth
d fallen wood,
hich creates open
rkland.

MOUNTAIN
COTTONTAIL

MOUNTAIN BLUEBIRD

GOLDEN-MANTLED
GROUND SQUIRREL
*Told from
hipmunk
y lack
of lines
on face*

LARKSPUR

ators will challenge the porcupine in direct combat. The mountain lion must first flip the animal over on its back so that its vulnerable belly is exposed to attack.

The open spaces at the edge of the ponderosa-pine forest are meeting places for many animals that use the forest as shelter during the daytime. Here there may be water, but there also is sure to be a more plentiful supply of grasses, shrubs, flowers, and other foods than is generally available in the forest itself. A community of mixed vegetation formed at a point where two different major ecosystems meet and overlap is known as an *ecotone*. The variety of living things in an ecotone may be greater than in the adjoining communities. This increase in the total number of species is known as the *edge* effect. Deer and other animals tend to congregate in these marginal areas of the ponderosa-pine forest. One of the shrubs grazed by mule deer and other game animals is the antelope brush or bitterbrush, also found in the chaparral and in the sagebrush flats. Another important food source for wildlife is manzanita, an evergreen shrub related to the very common kinnikin-nik, which has smooth reddish branches and produces yellowish berries.

The rocky forest floor beneath the ponderosa pines, dimly lit because of the shade cast by the broad crowns of the trees, is often bare of flowering plants. But in the intensely sunny open meadows, especially where moisture is plentiful, there are ever-changing displays of colorful blossoms. There are the beautiful blue-purple flowers of larkspur and monkshood; the dense, bright-yellow clusters of wallflower; the golden-yellow blossoms and beanlike pods of golden banner, or golden pea; the pink or purplish flowers and long seedpods of Frémont's geranium; and the yellow flowers and dark-blue berries of the low-growing, creeping evergreen shrub known as holly grape, or creeping mahonia.

The vast areas covered by the ponderosa-pine forest support many bird species. Among the more common permanent residents are the large, noisy Steller's jay, the only crested jay in the Rockies; the mountain chickadee, the only chickadee with a white eye-stripe;

the pygmy nuthatch, which lives almost exclusively in these forests and is constantly climbing the pines to pick insects from the bark of the trees; and in the summer, Grace's warbler, with its identifying yellow eyestripe. The black-capped chickadee frequents the more open areas. Among the more spectacular birds are the bright-yellow male western tanager with its red head, black wings, and two prominent wing bars, spotted in the summer, and the mountain bluebird, whose pure sky-blue covering is seen all year.

It takes about 150 years for the ponderosa pine to reach maturity. It is then that it achieves the imposing physique it can carry with dignity for an additional 150 years or more. At full growth, it becomes an important high-grade-lumber tree, extensively used in the production of numerous items for the home-construction industry. But throughout its lifetime, it faces attack. It begins its growth slowly, and as a seedling is subject to browsing by many animals, springtime frosts, and long seasons of drought. Fire is one of the hazards of the young tree's existence, and as an adult, it is not spared by insects. Trees are defoliated by butterflies and moths in their caterpillar stages. Beetles bore through the bark to eat the living tissue beneath. An entire grove can be destroyed by an infestation of these and other insects. The height of the trees attracts lightning, and there are many visible evidences of damaging strikes in the forests. Burnt-out areas are seized by aspen and lodgepole pine. Yet in all the western states, growths of ponderosa pine fill widespread mountain areas with an abundance that merits the name "ponderosa," meaning heavy in weight but also in the modern slang sense of "awesome."

ASPEN GROVE

Throughout the Montane Zone and extending into the Subalpine Zone, groves and thickets of quaking aspen stand out in delicate relief against the sedate pine, fir, and spruce of the Rocky Mountains. In the spring, the aspen leaf out in light green buds and produce great quantities of seed that fill the air with yellowish fluff. In the summer, the almost circular leaves with rounded marginal teeth, apple green above and pale silvery below, flutter in the slightest breeze, giving the tree its popular name—quaking aspen. The fluttering is caused by the angle at which the long, flexible, flattened stems are attached to the leaves. Within the grove, the shimmering sunlight through the leaves has the radiance of a mosaic, and from a distance, the mass of pale green sparkles against the dark green background of conifers. The brilliant red, yellow, and gold colors of the fall make an aspen grove one of the truly spectacular sights on the mountains, rivaling the riot of color produced each autumn by the trees of New England. In the winter, the slender trunks in their whitish bark, marked by black, warty scars, stand leafless against the icy-blue of the afternoon sky. No other deciduous tree is generally seen in the region of the coniferous forests, and few trees can make better claim to the title "a tree for all seasons."

In the Montane, and often the Subalpine Zone, stands of aspen may develop in areas that have suffered some drastic disturbance in the natural pattern of plant growth. This disturbance can be caused by erosion, rockslides, lumbering, open-pit mining, or disease. Fire is the most common cause, since lightning frequently strikes the tall ponderosa pine. In the absence of any such disturbance, plant life develops systematically. One community of plants, beginning with so-called pioneers, follows another in the process known as succession. *Succession* means a shift in the plant population of an area and a resulting shift in the herbivore and predator population. The process involves a gradual and predictable change in the composition of a natural community until an end point, called the *climax community*, is reached. This community consists of the most advanced plant life possible under existing climatic conditions in the area. Because a climax community can stifle competition by its ability to reproduce itself, it remains relatively stable unless there is a major change in the environment. Succession takes place everywhere. The area in which it takes place may be a few square feet or many square miles. The forces of change are both physical and biological. In each community these forces produce new environmental conditions. Under these changed conditions, new species of plants and animals thrive better than the old ones, and in the competition for living spaces, the new species displace the old. The rebuilding process therefore involves a methodical destruction of one community by the next higher community until the climax is reached.

Aspens are among the species of plants that cannot become established without at least a moderate amount of sunlight. In the mountain reaches of the Rockies, the climax community—depending upon local variations of temperature, rainfall, and soil conditions—is a forest of pine, spruce, or fir, or a combination of all three. Fire is a common disturbance in these regions, and when it destroys a stand of ponderosa pine and sometimes Douglas fir in the Montane Zone (or a forest of spruce and fir in the Subalpine Zone), it frequently frees old aspens from their inhibiting environment. Without competition from the overshadowing conifers, young aspens sprout in the moister areas from the roots of the old trees. These young trees grow quickly in the new sunlight. Soon any existing weeds and shrubs are crowded out, and an entire hillside or flatland is colonized by an aspen grove. Or the grove may have its beginnings from the billions and billions of aspen seeds that travel great distances in the springtime on their fluffy "parachutes." If these seeds find moist soil in an open, sunny area, they sprout quickly and in a few years send out their own roots from which other saplings can develop. Because many of the young sprouts originate from a single root system, an aspen grove usually has a uniform appearance in shape, bark texture, and branching pattern, all the trees flowering and leafing as one. Several hundred shoots can sometimes sprout from a single root system, so that large burnt-over areas are rapidly returned to forest land.

But the aspens are only in temporary possession of the field. They have relatively short lives, a hundred years or so, and since they require direct sunlight for growth, they cannot replace themselves in their own shadows. Their open pattern, however, permits just enough sunlight to filter through their crowns for slow-growing evergreen seedlings to begin recapturing their original territory. Year after year, the ground is made richer in humus by the decaying aspen leaves. The young evergreen trees begin to cast a deep shade. The air becomes cooler and more humid. The environment begins to change again. Over many decades, the evergreens will crowd out and suppress the aspens. Not all the aspens will die, though many of the survivors will be stunted. But a new evergreen forest will have been established. If fire or other disaster strikes again, the surviving aspens will repeat the entire process.

In the light shade of an aspen grove, especially in the flat areas irrigated by mountain streams and in the open meadows around the groves, ferns and wildflowers grow in profusion. One of the most interesting sights early in spring is the purple or violet cups of the pasqueflower pushing their way through the melting snow. Another springtime flower is showy loco. Its pink, pealike flowers on long stalks contrast with its silvery foliage. By June, the large white-and-blue blossoms of the blue columbine, the state flower of Colorado, make a cheerful display. This flower, with its slender spurs that extend backward from each of its five petals, is one of the more common wildflowers of the aspen groves. In July, the white, bell-shaped flower of the sego lily, the state flower of Utah, has made its appearance in the drier areas. The roots of this flower, about the size of a walnut, were used as food by the Indians. The entire summer is bright with a variety of colors: the deep blue of the mountain gentian, the golden yellow of the false lupine, the white clusters of pearly everlasting, the sparkling yellows of the sunflowers, the rosy pink of fireweed, and the brilliant red of Indian paintbrush.

Birds and other wildlife are also more plentiful in the summer in the flat aspen groves than in the heavy shade of evergreen forests. Hummingbirds can often be seen hovering over the blossoms, probing for nectar with their long, slender beaks. They are among the smallest of birds, measuring about three inches long and weighing only a fraction of an ounce. The broad-tailed hummingbirds are recognized by the metallic whistle of their wings, and their rounded tails and green crowns. The male is distinguished by its rose-colored throat. Two species of birds that nest in tree cavities are the tiny house wren, whose short tail is often cocked over its back, and the tree swallow, whose throat is white and upper parts metallic blue or green. Small parallel holes bored into a tree trunk are evidence of the presence of the yellow-bellied sapsucker. The sap that oozes from these holes is sucked up by the birds, and the insects attracted to the sap are additional food. Both male and female have red crowns, and the male has a red throat as well. Their backs are a mottled black and white, and their bellies are a dull yellow.

There can be no doubt about the presence of a beaver family. A dam across a stream, constructed of entangled sticks of aspen, cottonwood, and willows that have been plastered with mud, impounds the spring runoff to create a pond and acres and acres of irrigated fields. The pond and the surrounding land, if low enough, become the focal point of a varied and interdependent community. Marsh plants become established, frogs, snakes, and insects settle in, migrating birds rest near the water, and animals of the forest come to drink. Dead trees, drowned by the rising waters of the pond, provide homes and places of refuge for many of its inhabitants. The beaver's large, conical house at the edge

ASPEN GROVE

Young BEAR *climbed tree*

BEAVER

GLOBE ANEMONE

BROAD-TAILED
HUMMINGBIRD

COLORADO BLUE
COLUMBINE

TREE SWALLOW

ASPEN *with lightning strike*

BEAVER *gnawed from snowline*

ELK *or* MOOSE *bites*

BEAVER *gnawing*

YELLOW-BELLIED SAPSUCKER *drills holes, waits for sap and insects to fill them.*

SHOWY LOCO

GOLDEN SEGO LILY

"Clones" similar in bark, shape, branching patterns. They flower and leaf as one because they are all from same roots—"all are holding hands under the blanket" (same as GAMBEL OAK)—*a pioneer community in Montane (Canadian), Subalpine (Hudsonian) Zones on better soil than lodgepole pines.*

SEGO LILY

HOUSE WREN

of the pond is constructed in the stick-and-mud style of
the dam. The conical tree stumps around the pond
show the distinct channels of the beaver's tooth-marks
as it girdles a tree to be used as timber for building the
dam. The huge front teeth of the beaver never stop
growing. They are kept sharp by being rasped together
and honed down to a comfortable size only by being
constantly driven into tree trunks.

Repairing the dam, as well as the lodge, is an instinctive urge of the beaver. After it has eaten the bark of
the small branches, the remaining wood is interlaced
into the structures to give them added strength. Beavers
do not hibernate. In the fall, a supply of boughs and
small logs cut up into small sections is anchored in the
mud that collects behind the dam. This is the winter's
food cache and can easily be reached from the entrance
to the house, which opens below the frozen surface of
the pond. If the house were destroyed or abandoned
and the dam allowed to collapse, the pool and marsh-
land would be drained and the dependent plant and
animal community would disappear. Not all beavers are
architects and builders. Along especially swift streams,
some beavers will only dig into the bank from under
the water and burrow an upgraded tunnel leading into a
dry den. This may also be a survival technique designed
to avoid conspicuousness.

The rich fur of the beaver, used in hatmaking, inspired some of the early exploration of the continent,
and its value caused the beaver to be hunted almost
to extinction. Beavers were caught by placing a limb
smeared with beaver scent over a trap attached to a log.
When the beaver was caught, it dragged the log around
until it drowned from exhaustion. The trapper returned
to collect the log and his trophy. Beavers, as indicated
by their small eyes, are daytime animals; they nonetheless learned to work mainly at night when trapping was
at its height. This habit has apparently been passed on
to succeeding generations, so while the beaver hat is a
thing of the past, a daytime sighting of a beaver is a
rarity. Beavers are not as smart as they are reputed to

be. They do not fell trees so that they fall toward the pond. Most of the trees lean that way to begin with. There have been enough reports of beavers getting caught under the trees they were cutting down to suggest that most of their lumbering is instinctive. Today, beavers have been introduced into many protected areas, and the mountain stillness may at times be broken by the sharp snap of a beaver's tail on the surface of its pond. This is the warning of danger given to others just as the beaver dives for safety into the lower levels of its habitat.

There are markings on forest trees that indicate the presence of other animals, many of which leave identifying marks. What look like old initials are often the tooth-marks of elk, deer, or moose. These mammals do well most of the year because in summer they escape the heat and flies by moving up into high country. In winter they come down into the lower meadows to escape the extremely cold weather. Many of these areas, called "parks" or "holes," have become urbanized and are no longer available as winter ranges. The animals are now forced to subsist partly on aspen saplings, and the mature trees may show signs of damage from their foraging during other parts of the year as well. If the branches of an aspen appear to have been snapped or torn off, it is probably the work of an elk or moose browsing in the grove. If the bark of a tree has been split into shreds, the antlers of an animal have been scraped clean of their covering of fur, called "velvet." Aspen bark is said to be high in Vitamin K, the vitamin essential to blood-clotting, and it is believed by some that female elk nibble on the bark just before giving birth. Vertical scratches on the trunk of a tree are left by bears and bobcats when they sharpen their claws. A collection of young branches stripped of their bark or a scarred tree trunk may be the work of a hungry porcupine.

There are other agents of injury. The holes drilled by woodpeckers may become the entry point for disease. Engraver beetles burrow just under the bark, after killing a tree by severing its liquid-carrying tubes. The larvae of certain moths can completely defoliate a tree in its summertime fullness. The larvae of other moths feed on roots and young shoots, and can snip off a stem just above the ground. The tender bark of young shoots is a delicacy for the mountain cottontail, snowshoe hare, and meadow mouse. Despite all these depredations, the aspen survive, and as true pioneers, quickly reforest scarred areas with a cover of delicate beauty.

DOUGLAS-FIR FOREST

A Douglas-fir forest, like the ponderosa-pine parkland, is another of the climax communities of the Montane Zone. Generally, ponderosa pine will establish itself on the dry, south-facing slopes, while Douglas fir will occupy the moister, north-facing slopes, with occasional intermingling between them. In the higher elevations of the zone, there is a gradual change until the Douglas fir has taken over completely. In the montane forests on the west side of the Continental Divide, as soil, exposure, and weather conditions change, Douglas fir is the dominant tree, with only rare specimens of ponderosa pine. Along the streambanks of the cool canyons, white fir may appear in the southern and central regions of the West and blue spruce in the northern mountains, both sometimes also scattered among the Douglas Fir. In the southwestern mountains, the Douglas fir and ponderosa pine often merge to form a mixed forest.

The surest way to identify the Douglas fir is by its distinctive cones. These are reddish, three to four inches long, hang down from the branches, and have three-pronged bracts that project from between the scales. The bark on mature trees is thick and deeply furrowed in shades of dark and lighter brown. On young trees, the bark is smooth and gray, with numerous resin blisters. The needles are flat, soft, and short, with pointed tips that narrow at the base into a slender stem. In the Rocky Mountain regions, Douglas-fir trees usually attain a height of about one hundred feet, with trunks about two to three feet in diameter. The crown of the tree is in the shape of a cone, with branches that sweep down and then turn up at the ends.

Many species of trees in the pine family are subject to infestations of mistletoe. This is a parasitic plant whose sucker roots sink into the bark of the host tree, feed upon its sap, and ultimately deform or destroy the tree. There are various species of this plant, each of which attacks a different species of tree. These are not the mistletoes of Christmastime frolics.

The decaying plant material lying in the deep humus of a coniferous-forest floor provides the food for the plant called pinedrops. This plant lacks chlorophyll, the green coloring material with which ordinary plants, with the help of sunlight, manufacture their food supply. A plant that cannot make its own food but lives on and derives its nourishment from dead or decaying vegetable matter is called a saprophyte. The upright stems of pinedrops are reddish-brown, covered with glandular hairs, and support a series of downward-hanging, pale-yellow bell-like flowers. The plant dries to a darker brown in the fall, and the dead stalks stand up conspicuously on the forest floor, displaying their seed capsules into the following year. Another interesting and very distinctive plant is twinflower, a member of the honeysuckle family. This is a trailing evergreen, with very short upright stalks—not more than four inches tall—that grow in moist, shaded areas. The stalks divide into two at the top, bearing a pair of dainty, bell-shaped pink flowers that are quite fragrant. The plant is named for the way in which the blossoms are paired.

One of the characteristic game birds of the Douglas-fir forest is the blue grouse, a species commonly found throughout the region. The male has dark-gray plumage with flecks of black and brown, and an orange patch over each eye. The female has a mottled brownish appearance. Both birds seem to blend into their environment with maximum camouflaging effect; but the birds are surely there, for the dark canopy of the forest is often filled with the reverberating sound of their cries. In their courtship ritual, the male struts to attract the female, spreading its tail into a wide fan and inflating its purple neck sacs in full display. Courting, mating, and nesting take place in the spring and summer on the lower mountain slopes. In a reversal of the usual pattern of migration, these birds move up into the higher elevations during the winter months; there they feed on the needles and buds of Douglas fir and find shelter under its snow-laden branches.

If a small, short-tailed, short-legged bird is seen climbing headfirst down a tree, it is a nuthatch. In coniferous forests, the most common of these acrobatic birds is the red-breasted nuthatch. It shows a white stripe above its eye and a black stripe across the eye. Its upper parts are bluish-gray, and its underparts are a yellowish brown. Its principal foods are the small insects found on the bark of conifers and the seeds extracted from the cones. Its nest is usually built in tree cavities, and the entrances are often painted with pitch, apparently to discourage predators. Many of the owls in the forest are also cavity nesters. For purposes of field identification, owls are broadly classified into two groups: "eared" owls, which have tufts of feathers on the side of their heads resembling ears, and "earless," those that don't. The largest of the "eared" owls is the great horned owl, whose ear tufts are large and widely spaced and whose belly is barred with narrow horizontal stripes. The eyes of the owl are fixed in position, so that an owl's entire head must move as it shifts its field of vision. Most owls are nocturnal and begin to move about at dusk in their silent search for grouse, small mammals, beetles, and frogs.

Another common bird of the evergreen forest is the olive-sided flycatcher. It has a large bill, and its dark-olive sides are in sharp contrast to its white throat and the white streak down the center of its breast. It almost always perches on a dead branch in an exposed location on top of a tall tree. All flycatchers are small, have eye-rings and wing bars, and flip their tails in an up-and-down motion. They are usually seen in the open. Another bird of forest margins and clearings is Cooper's hawk, a powerful hunter that can maneuver efficiently because of its long tail and short, rounded wings. When not pursuing its prey, birds and small mammals, its short wingbeats are alternated with the graceful glides of all hawks.

COOPER'S HAWK

GREAT
HORNED OWL

BLUE GROUSE
Male

DOUGLA
SQUIRRE
*found
mostly
in Pacifi
Northwe*

TWINFLOWER

OLIVE-SIDED FLYCATCHER

RED-BREASTED
NUTHATCH

DOUGLAS-FIR FOREST

*Characteristic tree of Montane region
on western side of Continental Divide
in northern Colorado, Utah, and
Wyoming. Mixes with*
PONDEROSA PINE *elsewhere*

PINEDROPS

UE GROUSE
male

RED SQUIRREL, *or* CHICKAREE

Best sign of DOUGLAS-FIR *is pine cones at base with distinctive
three-pointed bracts jutting out from underneath scales.*

The forest is full of noisy little squirrels, often heard before they are seen. Their habit of gathering cones and burying them for later consumption is an important factor in the life cycle of the forest. For one reason or another, many of the stored cones are never uncovered by the squirrels. Seeds buried by these rodents germinate and come up more often than those lying haphazardly on the surface of the ground, especially since the most perfect cones are generally selected for these caches. The red squirrel, the smallest of the tree squirrels of the region, is one of these hoarders. It is usually seen sitting on a branch above ground, its reddish upper portion (with a black summertime line), its white belly and bushy tail set off by the green needles of the evergreens. Its constant activity does little harm and affords great pleasure to campers, hikers, and park visitors. A counterpart of the red squirrel is the Douglas squirrel, or chickaree. It is slightly darker than the red squirrel, and its belly is yellowish, not white. It, too, is an entertaining animal but is more abundant in the forest of the Pacific Northwest.

On the Pacific coast, where the winters are milder and the rainfall heavier than in the Rockies, the Douglas fir grows faster and taller, sometimes reaching a girth of eight feet and towering to a height of more than two hundred. This is the majestic formation that has fulfilled so many of the needs of the lumbering industry and given the Douglas fir the title of premier industrial tree of the world. Its size and its strong, durable wood account for its multiple uses, from construction timber to railroad ties, telephone poles, and structural beams, and to its foremost position in the manufacture of plywood. Douglas fir is commercially important wherever it grows.

Douglas fir, named after David Douglas, one of the early-nineteenth-century explorer-botanists, is not a true fir tree. One obvious difference is in the way the cones grow. In the true fir, the cones stand erect and disintegrate at maturity, leaving residual upright spikes on the branches. In a forest of Douglas fir the floor is littered with whole cones that can be easily identified by the three-pointed bract, which, like a little slip of brown paper, projects beyond each scale of a Douglas-fir cone. Nor is there any mistaking the deep furrows in the mastlike trunks and the shaggy appearance of the hanging twigs. The groves are cool and dimly lit, and a human visitor cannot help but be a little overawed by the strength, dignity, and permanence of the Douglas-fir forest.

SAGEBRUSH FLATS

The single most widespread plant in the West is sagebrush. Wherever there is dry soil, from basin floors to
the flanks of mountainsides, one of the species of this
plant may be the dominant vegetation. It is so pervasive
in the Great Basin Desert that this region is sometimes
called the Sagebrush Desert. Except on saline flats or the
highest peaks, the domain of this plant extends from
Canada to New Mexico and from the eastern Rockies to
the Sierras. When the cattleman first came to these
open spaces, nutritious grasses held their own against
the sagebrush. Both had grown together for thousands
of years. But overgrazing by cattle and sheep killed off
many of the grasses and permitted the less palatable
sagebrush to flourish. Today, miles and miles of valleys
and slopes are monotonously abundant with these low-
statured shrubs. Generally, the plant cover is not complete, but the bare places can sustain very little other
vegetation.

All of the species of sagebrush, whether tall or short,
have the same general appearance as big sagebrush, the
most common of all. This is a bushy gray-green shrub,
two to ten feet tall, whose wedge-shaped evergreen
leaves have three teeth at the ends and are covered with
silvery hairs. The individual plants look like small,
twisted trees, with large shreds of bark hanging from
gnarled trunks. They bloom pallidly in August and September, proclaiming the end of summer and filling the
air with a pollen that causes hay fever. The plant is well
adapted for survival in this arid and usually windy environment. It has a wide-spread, shallow root system
that can quickly trap rainwater over an area of many
square feet, and longer roots that search deeper down
for water. Its narrow leaves expose only a minimum
evaporating surface to the sun. The fine, silky hairs that
cover the leaves and account for their silvery sheen re-

tard the desiccating force of the winds. In addition, sagebrush protects its domain by inhibiting the growth of other plants. As they decay their fallen leaves discharge a compound that is toxic to competing vegetation. All in all, the big sagebrush is a faultless model of endurance.

Sagebrush should not be confused with other plants. It is neither the sage of Zane Grey's *Riders of the Purple Sage* nor the garden sage used as a cooking herb. It belongs to the family of composites, or sunflowers, and its common name *sage* is derived from the pungent, sage-like odor of its leaves when crushed. If the West has a distinctive smell, it is in the sagebrush flats after a rain, or around a campfire fed with its aromatic wood. Zane Grey's purple or desert sage is a member of the mint family. It has spine-tipped branches, silvery leaves, and bright blue to violet flowers, and often grows beside sagebrush. Though cattle eat sagebrush sparingly, it is nonetheless important to wildlife. It provides the staple food of the sage grouse and supplementary food for antelope, moose, mule deer, and elk. Its low, contorted form provides shade, shelter, and sites for nests and dens for a large number of the smaller animals of these parched open regions.

Another reserve food for the large mammals of the mountains is the shrub known as rabbit brush, a plant found growing among the sagebrush and in other places where the soil is poor or overgrazed. The taller form of this plant grows to a height of two to four feet, with very narrow leaves and slender branches covered with dense woolly hairs. It blooms late in the summer, and because its long taproots permit it to survive where little else will grow, its large clusters of yellow flower heads make a conspicuous display over extensive waste areas. The shorter form of rabbit brush is also a common Rocky Mountain shrub. Its narrow leaves are twisted, but its flower heads resemble those of the taller plant. Another shrub of the sagebrush domain is antelope brush, or bitterbrush. The species name of this plant, like that of big sagebrush, is *tridentata*, which describes the three teeth at the end of its wedge-shaped leaves. This shrub grows from two to ten feet tall and, in the spring, is solidly covered with yellow, roselike flowers. It is intensively grazed by livestock and game animals alike and, as a result of this pruning, often spreads low and hugs the ground. Its seeds are a favorite food of the smaller rodents.

Several beautiful wildflowers are to be found in the dry soil of sagebrush country. The most spectacular display is made by scarlet gilia, or skyrocket gilia, also known as the polecat plant because the odor of its crushed leaves is like that of a skunk. This plant often grows in extensive patches, and its numerous trumpet-shaped scarlet flowers contrast brilliantly with the backdrop of the grayish sagebrush. The plant is a biennial; a rosette of leaves develops one year and a flower stalk the following summer. Another common plant of the sagebrush flats is the sulfur flower, known as the umbrella plant because its small yellow flowers develop at the end of leafless stalks in a circular, spokelike pattern. Wyoming paintbrush, with its brilliant red flowers, is, like other paintbrushes, a partial parasite, extending its own roots until they pierce the roots of another plant, often sagebrush, and then siphoning off a supply of food from the host plant. The colorful tubes of blue penstemon and the golden, sunflowerlike heads of balsamroot are also products of the dry soils of valleys and hillsides. Many of these plants may often also be seen along the mountain roads.

The sagebrush flats shelter their own species of gallinaceous bird. This is the order that includes pheasants, turkeys, and grouse, as well as the common domestic fowl. The coloration of each species fits in best with the particular type of vegetation in which it lives. The blue grouse of the Douglas-fir forest blends imperceptibly into the coniferous growth. The earth colors of the prairie chicken are exquisitely tailored for the tall- and mixed-grass regions. The mottled grays and browns of the white-tailed ptarmigans are a suitable summertime match for their rocky mountaintops—in winter these colors are exchanged for a camouflage coat of white. The sage grouse is well concealed in its silvery-gray habitat. The annual courtship rites of these birds are often droll and colorful pageants; the males gather in the springtime on their dancing grounds, or leks, to challenge one another with impressive displays, ritualistic parades, and physical combat for the right to mate with the nearby females. While sage grouse may at times search out other plants and fruits, they rely almost exclusively, especially in winter, on the leaves of the sagebrush.

Two other birds are particularly associated with the sagebrush country. These are the sage thrasher and the sage sparrow, both of which generally build their nests in the sagebrush bushes and feed extensively on the supply of seeds produced by the various shrubs and grasses and on small fruits and insects. The marking of these birds make them fairly inconspicuous as they search for food under the cover of the shrubs.

Many of the small animals of the chaparral are well represented in the sagebrush, including lizards and snakes, and kangaroo rats, wood rats, squirrels, jackrabbits, pocket mice, and chipmunks. The pale colored sagebrush vole is confined to this habitat, digging its shallow burrows under the sagebrush roots and feeding voraciously on the sagebrush leaves. And wherever these animals thrive, the predators—the coyotes, gray foxes, bobcats, and badgers—are not far behind.

In late summer and fall, sparkling tangles of tumbleweed bounce across the open country until they pile up against the fences. The "tumbleweeds" include various annuals, the most common being the Russian thistle. This plant, not a true thistle, was introduced from Europe late in the nineteenth century and has since become a full-sized nuisance in many areas. It is greenstemmed, with stiff, prickly leaves and many branches, and takes on a roundish form. When dry and yellow, many of them are uprooted or break off at ground level and then roll before the winds, scattering their seeds over wider and wider stretches.

Though various plants and animals have adapted to the rather restrictive environment of the sagebrush, the treeless landscape has the severity and sameness of a

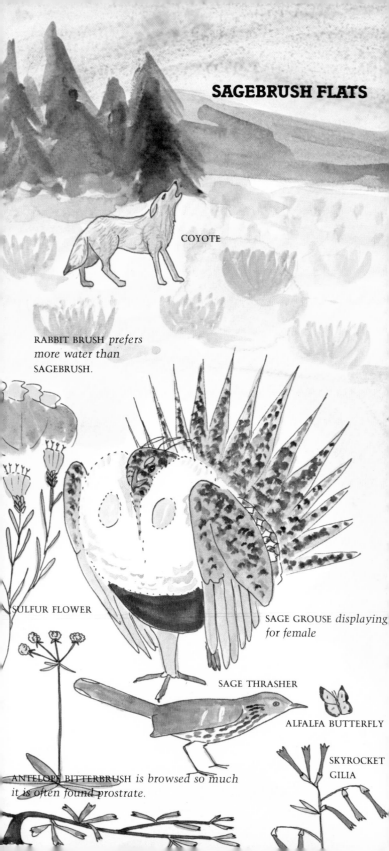

SAGEBRUSH FLATS

COYOTE

RABBIT BRUSH *prefers more water than* SAGEBRUSH.

SULFUR FLOWER

SAGE GROUSE *displaying for female*

SAGE THRASHER

ALFALFA BUTTERFLY

SKYROCKET GILIA

ANTELOPE BITTERBRUSH *is browsed so much it is often found prostrate.*

SAGE GROUSE

SAGEBRUSH

SAGE SPARROW

SAGE GROUSE

BALSAMROOT

WYOMING
PAINTBRUSH

BLUE
PENSTEMON

wasteland. As the sagebrush overtakes the native grasses, more and more acreage is needed to support the same number of livestock. To rebuild the grazing areas, large sections are now being deep-plowed or burned over, and planted back to grass. This may produce some temporary hardship for the wildlife, but will at least lend some softness and color to the vast spaces through which travelers now hurry just to get beyond them.

LODGEPOLE-PINE FOREST

If the stand of trees is so dense that practically nothing is growing on the ground, if the straw-colored trunks rise like spindly rods seventy-five feet or more into the air, and if dead trees are propped up by the branches of their neighbors because there is no room for them to fall to the ground, then you have come upon a lodgepole-pine forest. Passage through such a forest is further impeded by slim dead branches that cling awkwardly like snags to the trunks of living trees. Lodgepole pines are not very efficient at self-pruning, the process by which the lower branches of trees die and are discarded when there is insufficient light for their growth. Getting lost in these forests is not difficult. There are not many distinct landmarks, and the endless similarity of the poles can confuse even experienced backpackers.

The needles of lodgepole pine are dark green, stiff, and twisted, one to three inches long, fastened in bundles of two. The glossy cones remain closed and attached to the branches for several years, and each scale is tipped with a small prickle. Despite the height of the trees, their trunks are only from one to three feet in diameter. The bark of the tree is scaly, as if the slender poles had been rolled in brown cornflakes.

The common name "lodgepole" comes from the Native American practice of using the tall, slim trunks for making supporting poles for tepees and lodges. These were cut up into even lengths and allowed to dry in the summer so that they could be easily handled to set up winter camps. The trunks were also used, especially by the Plains Indians, to make travois, drag-sleds consisting of two poles framed and covered with a hide, and generally dragged by an animal.

Isolated patches of lodgepole pines in exposed windy areas become twisted into oddly bent shapes, which has earned for this species the Latin name *contorta*. The

LODGEPOLE PINE FOREST

PINE SISKIN

In Montane (Canadian) and Subalpine (Hudsonian). A pioneer community on dry, poor soil. Cones usually require fire to open. May be climax community in parts of Yellowstone.

RED CROSSBILL

PORCUPINE

SHORT-TAILED WEASEL (ERMINE)

PINEDROPS

RED-BACKED VOLE

GREAT GRAY OWL

HAIRY WOOD-PECKER

GRAY JAY

BEAR *scratching back and marking territory.*

"BEAR *tree*"

BLACK BEAR

SUBALPINE FIR *in understory will eventually take over as climax community.*

trunks of some trees are deformed by sharp crooks, called *knees*. A sapling overburdened by snow will bend at a right angle, and if not broken by the weight, will in time straighten itself out with another sharp upward turn. But in the crowded stands of a sheltered forest, the trunks grow straight and tall.

While lodgepole pines will grow in a wide range of conditions, the dominant forests are in the drier slopes of the Upper Montane Zone. The dry conditions, coupled with the density of the growth, the thin bark of the living trees, the tinder-dry wood of the dead trees and branches, and the highly inflammable resinous cones, make this type of forest extremely vulnerable to destructive fires. Yet the areas completely destroyed by fire recover quickly. This is because the heat of the fire will open cones that have remained closed for years, releasing seeds that germinate even in poor soil. Thousands of seedlings spring up at the same time. Since these young shoots not only tolerate but flourish in sunlight, a lodgepole-pine forest destroyed by fire will generally be followed by a new, close-packed forest with trees remarkably uniform in size. If the trees in such a forest are ten feet or so in height, people in the neighborhood can probably still remember the fire that cleared the land, possibly even how it started. For example, there is a small burn in Rocky Mountain National Park called the "Boy Scout burn of 1941," now covered with ten-foot lodgepole pines.

Fire seems essential to the existence of lodgepole-pine forests. Without it, these forests would be slowly replaced, particularly at higher elevations, by stands of spruce and fir, whose seedlings, tolerant of shade, would ultimately overwhelm the lodgepoles. Conditions that would be destructive to less vigorous trees do not deter the lodgepole pine. Often one fire follows another, and one forest of lodgepoles succeeds another. There are other pine trees that are also adapted to rebirth after a fire, but the extremely vulnerable lodgepole pine is the most proficient in checkmating its adversary. It is as if the lodgepole pine is willing to encourage its own destruction in order to preserve the genetic material of the species. Similar observations have been made about mammal species in which the behavioral characteristics of an individual animal are not in its own best interest but in the interest of the genetic material it is carrying. The thought is disturbing that there may be genetic material that has a mind of its own and will not accept the goals of its owner.

Lodgepole pines are also pioneers and may take over Douglas-fir and other forest areas destroyed by fire or by any other disturbance in the natural pattern of plant growth. In the moister, mountainous areas, such an area may be colonized by the sun-loving aspens. In the drier, poorer soils, a lodgepole-pine forest may be the successor, often extending into the Subalpine Zone. In recent years, man-made and lightning fires have been frequently controlled, with the result that many of the lodgepole stands have reached an advanced age. This has encouraged a phenomenal explosion in the number of insidious beetles that feed on the bark of the trees; together with the fungus they have brought with them, they kill whole groves and leave barren trees standing as "ghost forests." Fire control has upset the previous balance of the forest, and one enemy of the species has been replaced by another.

Much of this mountain country is black-bear territory. The smallest and the most widely distributed of the native bears, they are not always black. Their thick coats range in color from nearly black through brown and tan, and the lighter-colored animals often outnumber the black ones. The black bears do not have the shoulder hump or the dished face of the grizzly bear. They are often seen on roads and near campgrounds. They will eat almost anything, except forest plants, but will often damage trees when they sharpen their claws on the trunks or mark them with their territorial signs. They range through wide areas of the mountain country. They may venture up to the timberline looking for berries, anthills, beehives, and insects, and will often tear logs and stumps apart in the process. In the fall, they descend on the oak trees at lower elevations and feed heavily on the supply of acorns to build reserves of

fat for the winter dormancy period. Bears, unlike gophers and other smaller mammals, are not true hiberna-
tors, and on mild winter days will awaken from their
sleep and leave their dens in search of food. At such
times they will often claw the bark off the trunk of a
tree and leave their tooth-marks in the wood. Bear cubs
are born during the winter months and continue to
follow their mothers throughout the summer and into
the fall on their leisurely strolls through the forests.
By the end of the fall, the cubs are well grown and on
their own.

The trunks of the lodgepole pine may show traces of
other animals. The large tooth-marks of porcupines
sometimes completely girdle a tree, foreshadowing its
death. A red squirrel might make similar but smaller
marks. Often these marks are just above the height of
the snow level, indicating that much of this damage
occurs during the lean winter seasons.

One of the smallest mammals in the forest is the red-
backed vole. A rodent with long, soft fur, its head and
body measure about three to four inches and its tail,
one to two inches. It is distinguished from various other
species of voles by its gray sides and reddish back. Voles
are active during both the day and night and live primarily on the green forest vegetation. They in turn
become food for many predators, including other mammals and owls and hawks. One predator mammal is the
short-tailed weasel, which frequents colder regions than
its long-tailed cousin of the plains. It has a special adaptation for the colder and snowier climate: in summer,
its upper coat is dark brown and its underparts and feet
are white; in winter, it becomes entirely white. The tip
of its tail is black in all seasons. Its winter pelt is the
ermine of elegant fur coats. Like all weasels, it lives
primarily on rats, mice, and other small mammals, and
uses its canine teeth to pierce the backbone of its prey.
It serves a vital purpose in controlling the rodent population of the forest. Another predator of small mammals, especially of shrews and moles, is the mysterious
great gray owl, rarely seen because it spends most of its

time in the dense conifers of the upper elevations. It is a large, earless owl, with a face mask marked by gray circles, and a voice that fills the forest with deep, booming sounds.

A common year-round bird in the northern coniferous forest is the gray or Canada jay. It is sometimes called "camp robber" because it will enter campgrounds and persistently search out anything that is edible. A gray back, white throat, and white underparts are the markings of the adult. These jays mate for life and live in the same territory throughout the year. While they will eat almost anything, they store large supplies of food for the cold winter months. Their technique is to use their sticky saliva to bind together a collection of conifer seeds and buds, then to fasten the mixture to the branch of a tree for use when other food is not available. Another fairly common year-round bird in the pine woods is the red crossbill, named for the structure of its bill and the distinctive brick-red plumage of the male. They feed almost exclusively on the seeds of conifers, which they can extract quite easily with their crossed mandibles. The fieldmarks of yet another forest denizen, the pine siskin, include a sharp bill, notched tail, streaked breast, and patches of yellow on wings and tail. They often gather in large flocks and follow an undulating flight pattern. A rapid drumbeat on a hollow-sounding limb is the work of woodpeckers. These are the tree savers, moving around a tree trunk, supported by their stiff tails, and using their sharp bills and forked tongues to dig out harmful wood-boring insects. The common hairy woodpecker is black and white, with a long white stripe down its back. The male is distinguished by the red spot on the back of its head.

The pinedrops of the Douglas-fir forest are also to be seen in the decaying plant matter on the floor of the lodgepole-pine forest. Here, the shade is even deeper, so that little that depends upon sunlight can survive. Following a rain, there may be mushrooms of many colors. But the color of wildflowers will be seen only in the glades.

Inside the forest, the pine needles crack underfoot. The spindly trees that take a hundred years or more to reach maturity, perfectly uniform in all their dimensions, create an atmosphere of tedious monotony. There is no feeling of grandeur, but a grudging admiration for the vigor that can produce such a compact, orderly community over such vast areas of the mountains.

SUBALPINE FOREST

Beyond the range of the ponderosa pine, the Douglas fir, and the lodgepole pine is the highest, the snowiest, and the windiest of the forest zones—the Subalpine. Here Engelmann spruce and subalpine fir are the dominant species. The long-lived spruce is slow growing. The fir has a shorter life-span but is more prolific. Always found together, with the spruce much larger and more abundant, they form the climax forest in this zone. In the absence of some catastrophe, the forest will survive, for the seedlings of these trees are well suited to succeed their parents.

This subalpine forest, which begins at elevations of about 11,000 feet in the southern Rockies and at about 6,000 feet in the northern Rockies, covers vast areas of the upper mountainsides with a wide green mantle that is awesome in its sweep. If the uniform pattern is broken by an aspen grove or a lodgepole-pine forest, this is because fire or other disaster first destroyed the spruce-fir forest. In time, these areas will be reclaimed, since neither of the intruders can replace itself in its own shade, and the shade-tolerant spruce and fir will eventually overtake them. Close to the timberline, there is little competition, for in the thin soil of these high elevations, only the shallow-rooted spruces can take hold and survive. On the wind-beaten knolls and the dry, rocky outcrops, the twisted forms of limber pines and bristlecone pines embellish the landscape with their misshapen beauty.

The Engelmann spruce is a compact evergreen, more than one hundred feet in height, with a rich, reddish-brown trunk broken into large, loose scales. Its short, dense boughs sweep downward and accentuate its spire-like appearance. In the winter snow its heavy branches droop toward the ground like the flounces on the skirt of a mountain giantess. Its needles are single, often

SUBALPINE FOREST

Subalpine fir cones upright

Pitch

BLACK BEAR

MOOSE *"highlining" subalpine fir eats as high as it can reach*

Under snow, so MOOSE *can't reach*

MOUNTAIN LION

CANADA VIOLET

WILD PARSLEY

RICHARDSON'S GERANIUM

"Cranesbeak" seed of geranium

PINE GROSBEAK

ENGELMANN
SPRUCE *in
more sheltered
place than
subalpine fir,
perhaps
five hundred
years old*

*"Snow mat": Deep winter
snows press lower branches
down and they take root.*
SNOWSHOE HARE *takes refuge
in snow mat.*

SNOWSHOE HARE

RED
TWINBERRY

GOOSEBERRY CURRANT

BLACK TWINBERRY

MOUNTAIN LOVER

RUBBER BOA

HEART-LEAVED ARNICA

WOOD NYMPH

prickly, and are square in cross section. Its small cones hang down from the branches and are light brown in color, with thin, toothed scales. In the limited growing season of the high altitudes, it attains its lofty proportions only after hundreds of years. It is an important lumber source, even though its location makes harvesting difficult. In addition to being used for boards, what lumbermen call *saw timber*, it is used for telephone poles, railroad ties, and mine props.

The members of the fir family are also compact trees. Their needles are flat and soft, grow directly from their branches, and leave depressed scars on their twigs when they fall. Their cones grow upright on the upper branches and disintegrate in place at maturity, releasing their seeds and leaving candlelike centers affixed to the branches. The buds are often resinous, plump in shape with blunt tips. The firs are a source of paper pulp, and their slim, spirelike crowns make them ideal as Christmas and ornamental trees.

The needles of the subalpine fir are one to two inches long, with whitish lines on both sides. The cones, purplish at first, are two to four inches long, cylindrical and often sparkling with silvery beads of pitch. A mature tree has grooved, grayish-white bark and may grow from forty to seventy-five feet high. Its horizontal, whorled branches, too short and stiff to bend, carry the winter snow with the majestic poise of a mountain monarch. Some of the long, flexible lower branches, when pressed down by snow, may take root in the ground below. In the exposed areas at or near the timberline, subalpine fir and Engelmann spruce are among the trees that make up the strange, contorted forms of the so-called crooked-wood zone.

The spruce-fir zone is the range of the moose, the largest member of the deer family. There is no mistaking this mammal. The dark brown skin, overhanging snout, pendulous jaw flap, gray legs, and the massive, flattened, hand-shaped antlers of the male make for an ungainly body as it plods leisurely and noiselessly through an open area on its way to a lake, pond, or bog. Most of its summertime food is aquatic vegetation, which it browses while partly submerged in water. Its hooves spread open for a firm underwater grip. It is a good swimmer when it prefers the other side of the lake. Like all members of the deer family, the bull moose loses its antlers in early winter. Late in the summer it scrapes the velvet from the new pair by rubbing them against tree trunks, leaving the bark hanging in shreds. Now, for the rut, the antlers are fully wrought, ready for threatening display or bloody battle.

During the winter, when plants are under ice or snow, moose and deer must feed on bark, twigs, and saplings of evergreens and on the deciduous trees and shrubs in the valleys. To reach the tender branches, they will "ride down" a small tree or sapling by simply walking over it. Where there is a heavy snowfall, they can be found "highlining" a tree, eating the upper portions while the lower branches are buried underneath the snow. They will often move down to lower valleys where there is a smaller accumulation of snow. When caught in deep snowy areas, they become restricted in their movements and must depend on the trees and shrubs within a narrow range. If the snow persists, the hardship increases. The food supply shrinks, and the animals must reach higher and higher for the twigs and buds. Some of the smaller animals are weakened and do not survive. At times even the big bucks perish.

When herds of moose or deer are driven into the valleys by winter cold and snow, and food comes into short supply, the predators are not far behind. There may be coyotes, bobcats, and occasionally a pack of wolves. In the rugged mountainous regions, there may even be a mountain lion, the largest cat on the continent, known also by such other names as cougar, puma, and panther. These graceful predators still have a wide range, but their numbers have been so severely reduced by the constant pursuit of hunters and cattlemen that sighting one is a rarity. Their telltale signs are the long vertical scratches on the bark of trees made in the process of sharpening their claws. These are left behind in all seasons as these tawny, long-tailed cats pursue their prey soundlessly and continuously. Part of the year, the bull moose can fight back with his strong antlers, but in the winter, bulls and cows can defend themselves only with the sharp hooves on their powerful forelegs. When large mammals are not available, the mountain lion will take rodents, rabbits, and hares.

The hare of the Rocky Mountains is the snowshoe or varying hare. Both of its names derive from the ability of this large rabbitlike animal to navigate and survive in the winter. *Snowshoe* describes the oversize hind feet that make for ease in traveling across deep mountain snow. *Varying* refers to the difference in color between its seasonal coats. Its dark-brown fur is the perfect summer garment for hiding in the underbrush or among the roots of ancient spruce trees when the hare is not nibbling on the rich vegetation of the forest edge. In the wintertime, the dark-brown fur changes to white, blending with the scenery as the hare seeks cover beneath a snow-drift or under the low-hanging branch of a subalpine fir. Hares, like rabbits, store no food. In the winter they depend upon bark, twigs, and buds, and times of high population will often consume trees and shrubs down to the snowline. The hare selves are an important food for such flesh-eating tors as owls, minks, and foxes. Some of these predators have made adaptations for their own survival. The short-tailed weasel also changes from brown to winter white so that it can stalk its various prey with greater efficiency. The Canada lynx, a member of the cat family that feeds for the most part on snowshoe hares, has large feet for better mobility over deep snow. All of these animals are part of a timeless cycle in which abundance of prey is followed by an increase in the population of predators, and the scarcity of prey by a similar scarcity of predators. This phenomenon of the periodic rise and fall of specific animal populations has been studied for years, especially with respect to snowshoe hares and their predators, but while the cycle of increase and decrease has been confirmed, the precise environmental factors that trigger each phase still elude the scientists.

One of the two species of boas living in the West is

the rubber boa of the Rockies. Because of its grayish-green color, it is sometimes called the silver boa. It is a small, heavy-bodied snake, averaging less than two feet long, with a blunt head and a blunt tail. It is a proficient burrower and feeds mainly on small mammals, which it swallows whole. It seizes its victim with its teeth, coiling its body around the animal and then squeezing until death comes by suffocation. It is a nonpoisonous snake with a docile attitude toward humans. In danger, it will curl up into a ball, protecting its head but exposing its blunt tail, which it moves back and forth, simulating its head.

Pine grosbeaks are common summertime birds of the spruce-fir forest. There they live on the seeds of conifers and the fruits of plants on the forest edges. They flock and become nomadic in the winter, moving from one mixed woodland to another, where they feed on the buds of all trees if seeds are not available. The male is rose-colored with two white bars set against its dark wings. The female has similar wing bars, but is a dirty gray except for an olive tinge on its head and rump. A short, conical bill is characteristic of these and other seed eaters.

The flowering plants of the spruce-fir forest are species that prosper in a rich, moist environment. The floor of the forest is covered with a spongy mixture of decayed and decaying vegetable and animal material, generations of accumulated humus. Everywhere there is moisture, for this is the snow belt, where the winter accumulation lingers under the shelter of the evergreens and remains a source of meltwater well into summer.

In the less dense areas of the forest, enough sunlight penetrates to sustain a variety of flowers. Many of these provide food for the moose, deer, and elk. Richardson's geranium, which has a white or pale pink flower, is an abundant and widespread plant that makes spring and summer forage for these animals. Another widespread forage plant is the arnica, with its showy, bright yellow flower and heart-shaped leaves. So is the mountain clover, which is preferred as well by grouse and Canada geese. Grazers also find wild parsley palatable. Its white flowers grow, like Queen Anne's lace, in flat-topped clusters, and its pungent odor is like that of celery. The golden currant, or gooseberry currant, is also browsed by the large animals, while its juicy, dark-red or black berries are favorites of birds, rodents, and black bears. Two short and enchanting flowers of these moist woodland areas are the Canada violet, whose white petals have a tinge of purple on the reverse side, and the woodnymph, whose perfumed, waxy flower hangs bashfully by itself from its four-inch stem.

When we look up toward the mountain peaks, the spruce-fir forests appear to cover the scenic beauty with a continuous dark-green blanket. But all is not plunged in shade in these high regions. Here and there the forest comes down to the edge of a meadow, the view opens, and spread out in the sunlight is a dramatic display of sparkling wildflowers.

MOUNTAIN STREAM

A cold, bubbling stream rushes down the mountainside from its upland runoff sources of rain and melting snow. Its beginnings are tiny brooks or rivulets that join together until a permanent flow of water is established and soon is on its ceaseless move through the great river basins to the sea. The direction of the flow will depend on the origin of the first drops of water.

Running north and south through the Rockies, following a series of high peaks and ridges, is the Continental Divide. A drop of water that falls on the western slope of the divide ultimately runs into the Pacific Ocean or the Gulf of California; another drop that falls on the east side of the divide will flow into the Mississippi River or the Gulf of Mexico. Six of the great river systems of the United States begin in the Rockies. On the eastern slope, the Missouri, Arkansas, and Platte rivers flow into the Mississippi, while the Rio Grande empties into the Gulf of Mexico. On the western slope are the sources of the Columbia and its tributary the Snake, whose waters flow into the Pacific Ocean, and the Colorado and its tributary the Green River, which empty into the Gulf of California.

The long journey from the little stream on the top of the mountain to the delta at the edge of the continent is often expressed in terms of a life cycle of development—from infancy to youth, maturity, and old age. The infant stream is small, its bed is shallow, and its flow intermittent. In its youth, the stream becomes a raging torrent, with waterfalls, rapids, and deep pools, cutting downward into its channel as a V-shaped ravine is carved into the mountainside. At this stage, it is a fiercely powerful erosive agent, moving tons of boulders, rocks, and gravel in the process of leveling out the downward slopes of its bed. Gradually, the whitewater disappears, the downcutting is minimal, the velocity is

Shrubs by water

WILLOW

BIRCH

Seedpod stays on

ALDERS

Non-willows have several bud covers.

Bud capped by one covering

Seeds fall off

Thick tail

MOOS

RIVER OTTERS—*usually more than one*

TALL CHIMING BELLS

BROOK SAXIFRAGE

PINK-FLOWERED PYROLA
(BOG WINTERGREEN)

Nest of DIPPER

DIPPER
(OUZEL)

TIGER SALAMANDER

CUTTHROAT TROUT

FRESHWATER SHRIMP

BLUE-SPRUCE
eedles stiff and sharp

MOUNTAIN STREAM

Scales on cones
have smooth tips.

WHITE SPRUCE—
*needles on top
of branch.*

FRINGED PARNASSIA
(GRASS-OF-PARNASSUS)

BROOK CRESS

MINK

YELLOW MONKEY-FLOWER

SUBALPINE
BLUE VIOLET

NORTHERN WATER SHREW

RAINBOW TROUT

decreased, and the rate of flow is steady. By this time, the erosion is in a lateral instead of a downward direction, the river walls have become less steep, and the V-shaped valley of youth has become the U shape of maturity. In old age, the river moves slowly over a broad floodplain, around the bends and shifts in its course, building bars and clogging its channels with the sand and silt it keeps pushing in its current.

All along the stream are brightly colored birds. A belted kingfisher may be patrolling the area, moving on blue-gray wings from one exposed perch to another, watching for a chance to make its dive. It will swoop off, hover like a tern over its prey, dive headlong into the water, and come up with a fish, frog, or insect in its daggerlike bill. The western tanager in bright colors, the mountain chickadee with its white eye-stripe, and the pine siskin with its notched tail and yellow patches are overhead in the coniferous borders of the stream. If the banks of the stream are steep and sandy, they may be riddled with the nests of a colony of bank swallows, distinguished from other swallows by their narrow brown breastbands. Cavities in trees along the stream may be the nests of tree swallows, identified by their dark-green or blue backs and clear-white underparts.

The swiftest part of the mountain stream, where the water comes tumbling down in a torrent, is the domain of the dipper, or water ouzel—a curious bird that combines the qualities of a consummate singer and an expert underwater swimmer. It is usually heard before it is seen. Its sooty-gray body and brown head are quite inconspicuous, but its clear, ringing song can often be heard above the bubbling voices of the cascading water. When it is finally spotted, this small bird with a turned-up tail may be perched on a rock in the middle of the stream, bobbing and "dipping" its entire body, undisturbed by the surrounding turbulence. When it is ready to feed, it flies down into the foaming water and pokes among the rocks and stones for worms, bugs, snails, and similar choice tidbits. It is uniquely adapted for this fascinating performance. It has strong legs, an extra layer of down under its feathers to keep it warm, large oil glands with which to waterproof itself, and special features designed to keep water out of its nose and eyes. Using its partly opened wings as pushers, it walks along the bottom of the stream bed with apparent ease. With food in its bill, it flies back to a rock or log, shakes the excess water from its body, and returns to its bobbing, dipping, and singing. Even its nest is water-oriented. The round, bulky structure, woven of moss, grass, and pine needles, with an opening near the bottom, is set close to the stream bank, often behind a waterfall, so that its outside surface is constantly sprayed by the splashing current. The dipper remains in this misty habitat summer and winter, moving temporarily down the mountainside when the water becomes solidly frozen. It is a remarkable waterbird, at home only in the foaming, swift waters of a mountain stream.

The cold turbulence of a mountain stream makes it an ideal habitat for oxygen-loving trout, a precious prize of the freshwater fisherman. The turbulence drives in the oxygen and the cool water holds it in place. A stream that slows down and bakes in the sun will lose its oxygen and with it, its trout. Trout will also be driven from a stream when it becomes polluted with silt washed into it from the surrounding areas. Silt can slow down the current of a stream by raising its bottom, can coat the gills of fish and cause them to suffocate, and can clog their breeding grounds and bury the eggs before they mature.

Two species, the rainbow trout and the cutthroat trout, are endemic to the fast waters of the West. There are many subspecies of these trout to be found in mountain streams and lakes, and there is a form that enters saltwater and returns upstream to spawn. The two species hybridize and, in many streams in the Rockies, the so-called rainbow-cutthroat cross is not uncommon. Rainbow trout vary in color, but a mature fish has a distinct red lateral band. The cutthroat trout derives its name from the red markings on the outside of and below the lower jaw. Both species feed mainly on smaller fish, crustaceans, and insects. Many insects live on aquatic vegetation. In the slower parts of the stream some, like striders and treaders, are usually seen running about on the surface of the water. Brook trout, which does better in warmer waters, became established in the eastern mountains. Today, however, many species of trout are raised in hatcheries in massive numbers and released into suitable fishing areas, so that western species now inhabit eastern streams, and the brook trout and even the European brown trout survive in some Rocky Mountain waters where native competing forms are absent. The stocking of streams with aggressive newcomers often can result in the destruction of existing species unequipped to deal with dominant competitors. Rainbow and cutthroat trout, formidable predators, have had adverse effects on the more delicate introduced species. Other fish endemic to this region are the arctic grayling, a highly prized beauty with a large sail-like dorsal fin; the mountain whitefish, brown on top and silvery-white below, which competes with trout for food and space; and the Colorado squawfish, with a dusky-green back and silvery belly, that feeds voraciously on trout and other fish.

Along the upper mountain stream, alive with the flash of white water, where the dipper teeters at the edge of a waterfall and the rainbow trout tantalizes the fisherman, the blue spruce adds its murmur to the many sounds of the forest. The symmetry of its branches and the silvery-blue tint of its needles has made this the perfect ornamental tree. On the slopes of the Rockies, some are dark green, with blue showing only on the younger trees and on the new growth of the older trees. But many are entirely blue, adding a magical beauty to the stream banks. The blue pigment is not in the needles themselves but in the powdery substance that covers their surface and is especially noticeable in the summer. The needles are stiff, very sharply pointed, and rectangular or diamond-shaped in cross section. The cones are large and, as in the case of all spruce trees, hang downward. Another tree of ornamental quality found along the streams is the white spruce. The blue-green needles of this evergreen are twisted toward the upper side of the branches; its cones have thin scales with smooth, rounded margins.

The edge of a stream, at the lower elevations, is often crowded with deciduous trees and shrubs. Among the most common trees are the aspen and cottonwood. Many of the shrubs cling together in impenetrable thickets, the foliage of one mingling with that of its neighbor. Willows are widely distributed. Some are tall trees and others tiny creepers; most are medium-sized shrubs. Other shrubs that prefer the moist soil along streams are the birch and alder.

Many moisture-loving wildflowers brighten the wet, mossy edges and marshy meadows along the mountain streams. Fringed parnassus, or grass-of-parnassus, has a white, saucer-shaped flower, with conspicuously fringed petals. The basal leaves are kidney-shaped, and one small leaf clasps the middle of each slender stalk. Yellow or common monkey flower, which gets smaller in size at higher elevations, has bright yellow, snapdragonlike flowers spotted with red. Another widely distributed flower is the blue violet. The subalpine variety is short and compact, with pansylike, bluish-violet flowers. Chiming bell, or mountain bluebell, is a tall, leafy plant that often grows in large clumps; it is named for its loose clusters of drooping, bell-shaped, light-blue flowers. The tall, slender stem of brook saxifrage, found in moist, rocky areas, bears small white flowers and rises from a basal rosette of coarsely toothed, shiny green leaves.

The vegetation along the mountain stream changes as it flows from the high altitudes down into the lower valleys. But all along the stream, especially in early morning and evening, the mammals of the forest can be seen. Moose and deer come down to drink, river otter and mink leave their nearby dens to fish and play, and, among the smaller animals, the northern water shrew takes to the stream at night in search of small aquatic organisms to satisfy its enormous appetite. There are also plenty of amphibians nearby, noisy frogs and their voiceless relatives, the salamanders. The stream and the vegetation on its banks supply food and water to many of the animals that look for shelter in the adjacent woodlands. Increased animal activity is part of the "edge" effect that results when two communities, in a transition area known as an ecotone, meet to offer their joint resources to the species of both.

MOUNTAIN LAKE

Without the valley glaciers of the Ice Age, there would be few lakes in the Rockies. These were the glaciers that originated in mountain regions and developed into powerful streams of ice that sculpted the landscape as they moved downhill. They differed from the continental glaciers, which were massive sheets that covered the frigid zones and moved out in all directions toward the seacoasts. A valley glacier begins in a mountain hollow above the snowline, where the annual snowfall exceeds the annual melting and evaporation. As the snow collects it becomes compacted and is converted into a mass of ice. In time, the glacier begins to move out of the basin, carrying with it large blocks of ice loosened by frost and leaving behind a semicircular excavation known as a *cirque*. These cirques, the insignia of valley glaciation, became the beds of many of the present-day mountain lakes. They were filled with rain, snow, and meltwater when the climate warmed and the glaciers receded. Where cirques cut into a mountain on different sides, the peak was carved into the distinctive shape of a horn.

Other lake sites were sometimes carved along the path of a moving glacier. Rock fragments became firmly embedded in the ice, and as the glacier spilled over and began its slow downward march it shattered, abraded, and gouged the terrain over which it passed. V-shaped river courses were bulldozed and widened into U-shaped valleys. The greater the mass of the glacier, the more pronounced the shape of the U. Where the glacier met resistant rock, it moved laterally; where it was joined by a tributary glacier, it gouged deeper, creating a series of shallow depressions down the mountainside. When the ice disappeared, these hollows filled with drainage water and became the paternoster lakes, so called because together with the stream

PINTAIL

GADWALL

MALLARD

MOUNTAIN LAKE

MOOSE

TRUMPETER SWAN

TRUMPETER SWAN

AMERICAN COOT

GREEN-WINGED TEAL

WATER BUCKWHEAT
or LADY'S-FINGER

YELLOW POND-LILY

WIDGEON

BUFFLEHEAD

BARROW'S
GOLDENEYE

COMMON MERGANSER

RUDDY
DUCK

LESSER
SCAUP

YELLOW-HEADED
BLACKBIRD

REDWING
BLACKBIRD

BELTED
KINGFISHER

YELLOWTHROAT

RE'S
L

connecting them, they suggest the beads on a rosary. Another consequence of glaciation was the deposit of boulders, stones, and other debris left behind in the valleys by the melted ice. These morainal accumulations acted as dams and made ponds of many streams. The pulverizing action of the ice often produces a "rock flour," which gives glacier-fed lakes and streams the characteristic turquoise hue that is especially pronounced when blue sky is reflected in the water.

In geological terms the Ice Age ended only yesterday, and the lakes it produced will be gone by tomorrow. These gems of the mountains are doomed to destruction by built-in natural causes and, barring human or other interference, will disappear from the landscape. They will become the victims of the inexorable process of succession. In the discussion of the quaking aspen groves (see pages 74–82), it was noted that a site disturbed by fire or other catastrophe would, if left alone, ultimately revert to its former climax community. After the catastrophe, one group of plants and animals exists successfully for a time, but is gradually replaced by another that is better suited to the changed conditions. The process of change continues until a climax community is reached—a relatively stable environment capable of perpetuating itself. The type of progressive change that begins with a destroyed community is generally referred to as *secondary succession*. A primary succession begins on a barren site, free of any plant or animal life. A typical example is a collection of boulders in the alpine zone not yet colonized by lichens. Another example is a body of water in a rock or soil basin from which a glacier has only recently retreated.

The rate at which succession proceeds depends primarily on local conditions. The progression at lower altitudes is generally faster than in the cold, mountainous areas where plant activity is slowed down by rigorous living conditions. Nevertheless, even the high-country lake will in time be reduced to a pond, then to a marsh, followed by a meadow, and in the end by a forest. In the early stages, the overflow from the lake erodes the outlet, so that the capacity of the basin for holding water is gradually reduced. In addition, the basin itself is generally not watertight, thus offsetting the supply from surface streams. As these streams empty into the lake, loads of debris are deposited as sediment on the floor of the basin. This begins to decay, stimulating the growth of algae. The algae become the basic foods for plant-eating animals, which in turn become food for flesh-eating animals. The soil and water are enriched.

Plants begin to emerge at the edge of the water. Various species of new plants begin to float and accumulate on the surface of the water. Underwater plants begin to develop and cover the lake bottom, raising the level of the bottom bit by bit. The vegetation on the perimeter of the water is constantly moving inward, reducing the size of the body of water little by little. Wildflowers that can tolerate wet environments begin to colonize the perimeter. The imperceptible march into the water is carried on by a veritable phalanx of plants, generally in three ranks: the low grasses and sedges, the middle-sized shrubs, and the largest of living things, the trees. All these plants and a variety of vines contribute their leaves, stems, twigs, and roots to a ceaseless process of decay that builds up a rich soil close to or above the surface of the water. This permits the phalanx to move forward. In the end, perhaps after thousands of years, the last vestige of the lake is gone. The transition is complete as nature's technique of land reclamation reaches its climax.

Though mountain lakes are involved in the long-term successional process, they all have relatively stable communities that show little change decade after decade. Those at higher elevations, especially where the basins are in sterile bedrock, are too cold most of the year to support even small plant and animal organisms in large quantities. Their clear waters, icebound for much of the year, are not very hospitable to living things. At lower altitudes, where succession has already produced some aquatic and shoreline vegetation, the plant and animal community is varied and visible.

Vegetation in lakes and ponds is generally classified

on the basis of zones in which the plants live: the floating zone, the emergent zone, the submersed zone, and the border zone. One of the most interesting plants in the floating zone is the insect-eating bladderwort. This plant is supported on the surface of the water by little inflated bladders that have a trigger system of fibers. When a small organism swims against the sensitive hairs on the bladder, a trap door opens, the organism is sucked inside by the rushing water, and the door closes, all in a fraction of a second. The victim is then digested by enzymes in the plant. The leaves of the bladderwort are finely divided into feathery segments, and its small yellow flowers finish in a curving spur. Another floating plant is the minute perennial known as duckweed. It is generally no more than a quarter of an inch long, but these egg-shaped discs, lacking true leaves and stems, grow so profusely that they often form dense floating mats. Its flowers are almost microscopic, but these are rarely produced. Reproduction is mainly vegetative—by the formation of a bud that separates to start a new plant. As indicated by its name, the plant is an important food for ducks and a favorite of other waterfowl as well.

The emergent plants are those that are rooted in the soil and grow through the water. These are the most striking plants in the lakes and ponds. The tallest is the common or broad-leaved cattail, with leaves often six feet high. The flowers are in a cylindrical spike; the dense, brown lower portion is the female flower, and the upper portion is the male flower. While cattails have some limited wildlife food value, they serve best as roosting places and nesting sites for many birds and waterfowl. An emergent plant in the colder regions is the yellow pond lily. Its long stalks rise from roots embedded in the mud. These support waxy, yellow, cup-shaped flowers and large, round or heart-shaped, leathery leaves that float on the surface of the water or just below it. The leaves provide an excellent hiding place for fish. The seeds, called *wokas* by the Indians, were ground into a flour or roasted and eaten like popcorn. The seeds are a food for ducks, and the leaves, for moose. Water buckwheat is another plant with oval, floating leaves. It prefers shallower waters than those in which the pond lily thrives. The long stems lie prostrate on the water and end in erect, bright-pink flower clusters. The seeds are also an important food for ducks and other waterfowl.

The submersed plants are those that live beneath the water. A basic characteristic of these plants is that they are finely divided into long, stringlike branches and leaves. Pondweeds, which grow up to three feet tall, make up a large family of these plants. Their underwater growths are so dense they sometimes interfere with swimming and boating. They furnish good cover for insects, crustaceans, and other small animals, which in turn provide an excellent habitat for fish. Pondweeds are among the most valuable foods for many waterfowl, which eat the tubers and rootstocks, as well as the stem, leaves, and seeds. Mare's tail, a member of the water milfoil family, has short leaves in whorls of six or more. The plant usually grows entirely underwater, but at times a part stretches above the surface. The leaves are limp when submersed and rigid when emergent. Water buttercup, a plant often mistaken for pondweed and water milfoil because of its finely dissected leaves and branched stems, also grows in dense patches. Its leaves and stems are submersed, but its small white flowers generally extend above the water.

Plants in the border zone are generally grasses and related plants. Bur reeds grow in the shallow waters or in the adjoining wet soils. These plants have long slender leaves, with clusters of white flowers, mainly at the side of the stems, that ripen into burlike seed-heads. The seeds are eaten by water birds, and the entire plant provides food for muskrats and sometimes for deer and moose. Other grasslike plants in the shallow waters may be members of the sedge family, many of which grow in colder regions. Most sedges have solid, jointless, triangular stems with three-ranked leaves, as compared with true grasses, which have hollow, jointed, round stems with two-ranked leaves. Various species of this family grow abundantly in the wet areas of the Subalpine and Alpine Zones. One member of the sedge family seen in the colder regions along watery borders is cotton sedge, conspicuous at maturity because of the distinctive cottony growth of its flower heads. The round-stemmed giant bulrush grows at lower elevations in dense patches over large areas, but rarely reaches more than five feet tall in the mountains. Plants in the border zone furnish nesting cover for warblers, black-birds, coots, ducks, and geese, and their seeds are eaten by many of these birds. Some of the plants provide food for muskrats, moose, deer, and pronghorn.

The variety of animal life in and around a mountain lake generally increases as the lake moves along in the process of succession. At colder elevations where lake bottoms have little soil and are still primarily bedrock and where prolonged freezing weather discourages most growth, plant and animal life is minimal. But where a lake has advanced to the point where there is a good supply of aquatic and shoreline vegetation, a food web has been established that includes microscopic plants and animals, insects, worms, crustaceans, mollusks, am-phibians, snakes, fish, birds, and mammals. Here are the sites of great scenic beauty, where elk, deer, and moose come down to drink and browse, where there is ample food for ducks and geese, and where the deep voice of the trumpeter swan, once close to extinction, may at times still be heard.

SUBALPINE MEADOWS

A subalpine meadow is just one act in the long-term drama called *succession*, which begins with a lake, which in time is reduced to a pond, then changes to a marsh, followed by a meadow, and concludes with the finale of a climax forest. The stage set for the summer-meadow scene in this drama is the highlight of the pageant. The stage is covered with a wide carpet of grasses and broad expanses of brilliant wildflowers. The backdrop is a majestic tapestry of the rich greens of pine, fir, and spruce, setting off the apple-green leaves and the whitish trunks of shimmering aspens. Off to one side is a marshy area fed by a stream or a trickle of a creek. On the other side, where the meadow reaches up to the surrounding forest, tangled shrubs spread out in straggly thickets. Many of the performers just have walk-on parts, making occasional visits from the neighboring mountains. One interesting group in the permanent cast is the widespread rodent population, some of whose members rarely appear aboveground.

Where the soil is slightly moist and easy to work, hardly any meadow is without the telltale signs of the pocket gopher. In the wintertime, when the meadow seems dormant, the pocket gopher is still at his tunnel-digging activity, pushing excess dirt up into the snow. These plugs and long cores of earth stay frozen until the late thaw, when they settle slowly to the ground as long ropes of dirt that sometimes crisscross over each other. Summer tourists who stumble across these "mine tailings" sometimes mistakenly assume that they are hollow tunnels. The pocket gopher is a subterranean creature, feeding from below on bulbs, tubers, and fibrous roots. At times it comes to the surface to forage, but will often pull a particularly desirable plant down into its tunnel by the roots to enjoy the leaves, blossoms, and all. Their endless digging is important in the

129

SUBALPINE MEADOW

COYOTE

SHOWY GREEN
GENTIAN

CALLIOPE
HUMMINGBIRD
*prefers red
flowers*

FALSE
HELLEBORE

INDIAN
PAINTBRUSH

SILVERY
LUPINE

PINK PLUMES

PERRY'S
HAREBELL

GOPHER *tunnel sign*

MOUNTAIN
VOLE

YELLOW
EVENING-PRIMROSE

NORTHERN
POCKET GOPHER

FIREWEED

WOOD LILY

FLEABANE

BEARD TONGUE

LEAFY-BRACT
ASTER

CLUSTERED
PENSTEMON

WESTERN JUMPING MOUSE
*eats grass by cutting
stems of tall plants
in short lengths
until it gets to the
seeds.*

PARRY
GENTIAN

WHITE-TAILED
PRAIRIE DOG

HEART-LEAVED
BUTTERCUP

soil-building process: it brings subsoil to the surface and continuously aerates the earth.

The *pocket* part of the gopher's name refers to the external, fur-lined cheek pouches that open on both sides of its mouth. They are truly buck-toothed, since their orange incisor teeth are exposed even when their mouths are closed. Fur extends behind their incisors between the teeth and the lips. In this way they can carry rocks and dirt in their front teeth while keeping their lips tightly shut. Their tunnels are quite narrow, with little rooms off to the sides for nurseries, bedrooms, and toilets. They are not hibernating animals; a supply of food is kept in cold storage all winter long.

Another interesting rodent is the little western jumping mouse, which spends eight or nine months of the year in hibernation in a tunnel it digs three feet below ground. During the one third to one fourth of its life when it is actually "alive," in the sense of moving around, it lives mainly on seeds found near streams in mountain meadows. It has the unusual habit of pulling down a tall plant and in the process neatly cutting it into segments, which it organizes into little piles of grass stems. Most of its foraging is done at night, and it is seen during the day mainly when its nest, hidden on the ground under the protection of herbs and grasses, has been disturbed. It has long hind legs, somewhat like the kangaroo rat of the desert, a long tail, and a dark-brown color above, which becomes pure white below. Unlike the kangaroo rat, which lands after its jump on just its hind feet, the jumping mouse lands on all fours, since it has better developed forelegs.

In the wintertime, when the pocket gopher is active underground and the jumping mouse is hibernating in its burrow, white-footed deer mice and meadow voles emerge from their tunnels and leave tracks in the snow as they forage for food. Their nests are under the snow, built of dried grass. During the warm months, meadow voles construct underground nests and runways in the grass. When their population is high, the intricate network of these runways is plainly visible.

There are many species of voles in North America,

nearly a hundred, ranging from five to ten inches in length, all stoutly built, with ears embedded in rather long fur. As is true of a number of rodents, the vole population fluctuates from year to year with cycles of abundance and scarcity. Every environment has a "carrying capacity" for its resident species. This capacity is the amount of essentials, such as space and food, that it can supply to its inhabitants. When this capacity is exceeded, either the birth rate must decrease or the death or emigration rate must increase. The volelike lemmings, which rely on emigration for their solution (which ultimately increases the death rate), have the best-known "boom and bust" cycle. Their doomed pilgrimages across the northern tundra, though highly spectacular, have not sufficiently dramatized the need of many other animals, including humans, to contend with the carrying capacity of their respective environments. In many of the national parks, the carrying capacity for visitors is now being considered, and as a result, some previously popular hiking trails are periodically closed. There are some areas where birds are less endangered by natural causes than by bird watchers eager to add the sighting of a rare species to their life lists. Many of the voles have learned to adjust their birth rate when their numbers become excessive. Fetal and newborn mortality also increase markedly. Humans are perhaps the only animals that consciously strive to increase the carrying capacity of specific areas. On farms chemical fertilizers are brought in from the outside. In cities bigger and bigger buildings are erected. But slowly limits are being reached. There was a time when lemminglike emigration was a popular human solution. "Go west, young man," an expression used over a hundred years ago by Horace Greeley in the New York *Tribune*, has now been replaced by R. Buckminster Fuller's "Spaceship Earth."

Unlike the black-tailed prairie dog of the wide plains, whose colonies originally spread over many miles and numbered in the thousands, its cousin the white-tailed prairie dog had to limit its numbers because of the lower carrying capacity of the mountain meadows it

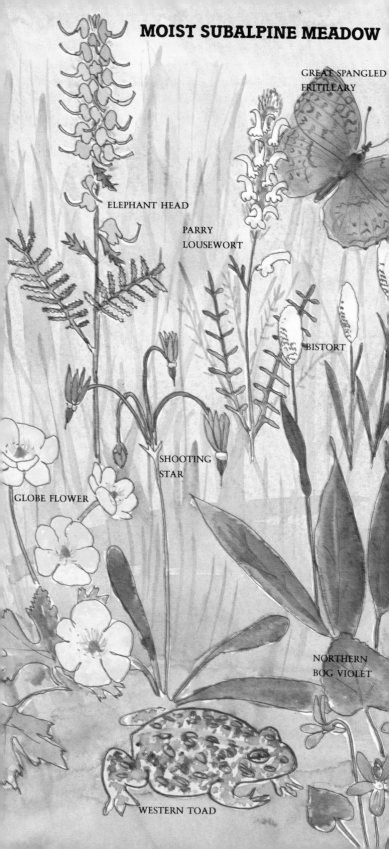

MOIST SUBALPINE MEADOW

GREAT SPANGLED
FRITILLARY

ELEPHANT HEAD

PARRY
LOUSEWORT

BISTORT

SHOOTING
STAR

GLOBE FLOWER

NORTHERN
BOG VIOLET

WESTERN TOAD

WEIDEMEYER'S
ADMIRAL

MONKSHOOD

JACOB'S
LADDER

WESTERN GARTER SNAKE

ROCKY MOUNTAIN
FRINGED GENTIAN

MARSH MARIGOLD

calls home. The plains species must contend with periodic floods, so it builds its burrow with a large cone of dirt, which serves as a dike to keep the entrance above water level and doubles as a lookout tower. The mountain species has no need for such protection against flooding, and so its mound is usually less carefully prepared, with the dirt just pushed off to one side. Because its colonies are smaller, each white-tailed prairie dog must be especially alert for danger signals, since it is less likely to receive a warning from another member of its group. Perhaps to compensate for their lack of numbers, they are very gregarious and helpful to each other. In Bryce Canyon National Park, where white-tailed prairie dogs have been reintroduced to a meadow near the road, park rangers have discovered that they must remove any prairie dog that has been hit by a car. Otherwise its relatives will be found on the highway endangering themselves in an effort to assist the unfortunate member of their colony.

As compared with the very elaborate burrows of the plains prairie dog, the tunnels of the mountain species go more or less straight down into the earth, with side tunnels at right angles. For a while it was thought that the vertical shaft was dug until water was reached, but this apparently is not the case. Its juicy diet of grass and roots supplies the mountain species with most of the water it needs. It may also be able to extract water from the starches in the seeds it eats in late summer. The tunnels go just deep enough for comfortable hibernation in the freezing winter weather. In the spring, after the young seem to be able to care for themselves, the parents often build a new burrow at some distance from the old one. Depending on one's viewpoint, they are either sagacious parents or deserters. The young, however, soon put on a thick coat of fat and by mid-October are ready for the winter's hibernation.

Whatever the fascination of the diverse wildlife activity in the mountain meadows, the most striking feature for the summertime visitor is unquestionably the breathtaking show of wildflowers. This is the great appeal of the subalpine areas, whether in the Alps or in the Rockies. Although there is often a spectacular display at the lower elevations, beginning in April in the southern deserts and going through June in the ponderosa parkland, these regions are dry with but few blossoms in midsummer when most vacationers arrive. But in the subalpine meadows, this is the period of peak blooming, when the high mountains display numerous jewels in their crowns.

For many people, the enjoyment in viewing wildflowers increases with the ability to identify them by name. This depends primarily on being able to recognize the differences in structure between one species and another. For the amateur, the emphasis for this purpose is usually placed only on the color. The subalpine meadow is an ideal place for even the amateur to take a closer look at plant structures and to begin to ask more questions. To do this, it is not necessary to discuss plants scientifically, but a few technical terms are essential for communicating the differences between wildflowers. A knowledge of these will not transform the viewer into a botanist, but will provide a basic vocabulary and may in time reduce the effort needed to know wildflowers and call them by names.

To begin with, a flower is more than a splash of color. A complete flower is made up of four sets of parts, each with a job to perform, and all organized for the essential purpose of reproducing the plant. The four sets of parts are attached to a special stalk called a *pedicel* at a point called the *receptacle*. The first, or outermost, set is the

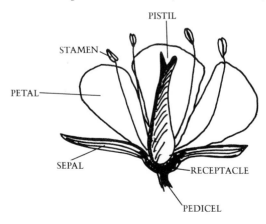

calyx. Its parts are the *sepals*, which are usually green in color. Above the calyx is the second set, the *corolla*. Its parts are the petals, the most colorful portion of the flower. Next come the *stamens*; bearing the pollen, these function as the male element in fertilization. The final set is the *pistils*, which contain the *ovules*. The pistils develop into fruits and the ovules into the seeds inside the fruits.

Now it may be helpful to notice how the flower itself is attached to the rest of the plant. This is more complicated than it seems because not all flowers simply sit at the end of a stalk like the wood lily, a beautiful flower that used to be seen widely but is now becoming rare because it has been picked so frequently. In some species, the individual flowers are combined on the same main stalk in a variety of ways. Fireweed, a plant of disturbed ground that is quite common by the sides of roads as well as in areas that have recently suffered from fire, has a main stalk called a *peduncle* to which separate flowers are attached by their own stalks. Such an arrangement is called a *raceme*. Elephant head, whose flowers have a long, twisted beak that resembles an elephant's trunk, are attached directly to the stalk without little stalks of their own. This arrangement is called a *spike*. A compound raceme is called a *panicle*. The tiny flowers of grasses are often attached to racemes that are themselves attached to a primary main stalk. Flat-topped flowers, like the well-known Queen Anne's lace, are divided into three types. There are the *umbels*, in which the stalks originate from a main stalk like the wires of an umbrella, all from one main point. A *corymb* is a flat-topped flower whose stalks arise from different points along the main stem. A *cyme* is a flat-topped flower cluster that blooms from the center toward the edges. It therefore often consists of some open flowers and some buds.

All of this indicates that the position of flowers on a stalk has been modified in various ways during the long process of evolution. Perhaps a tight group of flowers is a more attractive target for insect pollination than separated flowers. One way for a number of flowers to get

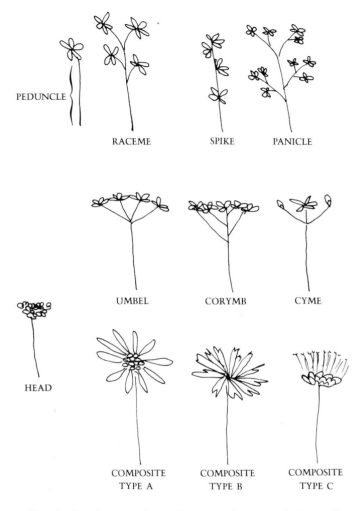

PEDUNCLE

RACEME   SPIKE   PANICLE

UMBEL   CORYMB   CYME

HEAD

COMPOSITE TYPE A   COMPOSITE TYPE B   COMPOSITE TYPE C

together is for them to bunch up at the top of the stalk. A *head*, as in the case of a clover, is just such an arrangement. It is quite easy to see the individual flowers, but they are so close together they form a unique little ball. A much more advanced development occurs in the case of the *composites*, which, as the name suggests, describe tight arrangements of component flowers. To the uninformed observer, an aster, a fleabane, or a sunflower may appear to be a single flower rather than a group of flowers. In this case of composites, the "flowers" are composed of an inner disc and an outer ring of what seem to be petals. In fact, each of these "petals" is a complete flower, called a *ray flower*. Likewise, the inner portion is composed of a tight group of *disc flowers*. Fleabanes as a rule have many more ray flowers than asters, but, together with many other composites, they make up a group that have both disc and ray flowers combined to appear like one big, intricate blossom. There are two other groups of composites. Dandelions, hawkweeds, and others have only ray flowers. Thistles, pussy-toes, and others have only disc flowers.

Another obvious structural feature of a flowering plant is the arrangement of its leaves. Leaves that cluster at the base of the flower are called *basal*. If each leaf has a partner at the very same level on the stem but on the other side, the arrangement is called *opposite*. If the leaves move up the stem one at a time, they are *alternate*. Jacob's ladder, one species of which grows under the wind timber of the treeline and another in subalpine meadows, is named for the neatly arranged "ladder" of paired leaves that climbs up its stem. Three or more leaves may be *whorled* around a stem. All these are called simple leaves. Some are *petioled*; others *sessile*, *clasping*, or *perfoliate*. Compound leaves are those in which several leaflets combine to make one large leaf. A leaflet, as distinguished from a leaf, has no bud at its base from which a new leaf may emerge. Leaflets arranged like fingers around a central point are called *palmately compound*. Those that are arranged on either side of a central stem, like a feather, are called *pinnately compound*.

The margins and shapes of leaves are important. The more common margins are those that are *entire*, *toothed*, *lobed*, or *dissected*. Among the various shapes are the *heart-shaped*, *ovate*, *lanceolate*, *elliptical*, and *linear*.

The shape of the flower itself is a significant structural feature. There are flowers that are symmetrical in the round, *radially symmetrical*, and those that are symmetrical only left to right. Globeflower is radially symmetrical, while monkshood is *bilaterally symmetrical*. Among the radially symmetrical flowers, there are those whose petals are free and those whose petals are

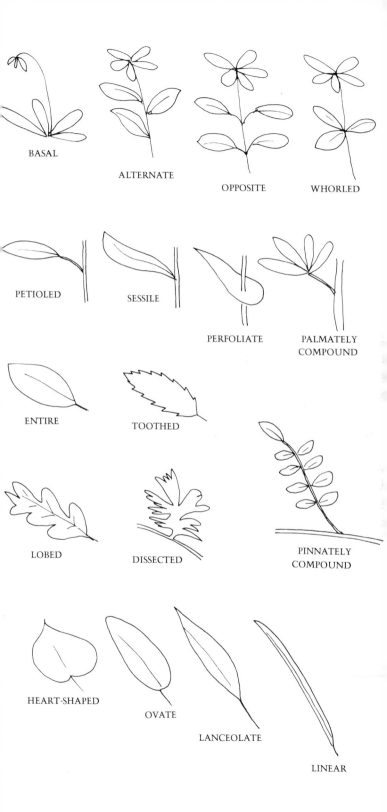

BASAL

ALTERNATE

OPPOSITE

WHORLED

PETIOLED

SESSILE

PERFOLIATE

PALMATELY COMPOUND

ENTIRE

TOOTHED

LOBED

DISSECTED

PINNATELY COMPOUND

HEART-SHAPED

OVATE

LANCEOLATE

LINEAR

RADIALLY SYMMETRICAL   BILATERALLY SYMMETRICAL

connected to form a shape like a dish, a bell, a tube, or a funnel. Among the bilaterally symmetrical flowers, there are quite a few very distinct shapes like the orchids, the spurred violets, the lilies, and the iris.

In some cases there are special identifying features. The penstemons, or beardtongues, have four regular stamens and one hairy ("bearded") stamen with a tuft or brush of hairs at its tip. Penstemons are numerous and showy and easily identified by their two-lipped bilaterally symmetrical tubelike flower with this special stamen. Another dissimilarity may involve the position of the ovary. The ovary may be *superior*—that is, above the point of attachment of the petals; or it may be *inferior*, below this point of attachment. Each flower distinguishes itself by some modification of the overall scheme of the plant kingdom.

Most of these structural characteristics are fairly visible and easily understood, and even a limited knowledge of the relevant vocabulary can destroy the fear of botanical explanations and result in a quantum leap in your knowledge and enjoyment of the wildflowers that bloom in the sunlight of the high mountain meadows.

TALUS ROCKSLIDE

High up on cliffs, generally not far from the timberline, great piles of rocks of all sizes and shapes seem to be clinging to the mountainsides. These are the product of the slow but inevitable breakdown of outcrops and peaks as weathering forces reduce and decompose the rock masses and move the particles and fragments toward their ultimate destination—the sea. These destructive forces include the weakening of rock surfaces by changes in temperature, the abrasive action of wind-blown sand and hail, the dissolution caused by atmospheric gases, the widening of crevices by the roots of plants, the erosive action of perennial rain, the grinding and scouring of glacial ice, the chemical action of acids released by lichens, and the expansive power of freezing water. Some of these forces cause a grain-by-grain disintegration of the rock surfaces. Others result in the peeling of slabs or flakes from these surfaces in a process known as exfoliation.

The largest fragments of rock debris are generally produced by the action of freezing water in the network of joints in the bedrock. These joints, fractures caused by natural strains and weaknesses, allow the free entry of water. The freezing of the water trapped in these fractures has a tremendous bursting force, splitting open the small crevices and producing the large-sized rubble, especially in the colder regions where the thawing and freezing action is repeated over and over again. Gravity, aided by wind, rain, and snow melt, begins to pull the debris down the mountainside. Small fragments move faster than large blocks. Much, of course, depends on the angle of the slope. The descent on a gentle slope may take so long that the material may be whittled down and pulverized before it reaches the bottom. On a steep slope, large fragments may descend quickly, before they can be reduced in size, and get caught in a natural depression or come to rest on a bench at the base of the cliff. These sloping accumulations of jumbled rocks are known as talus or scree.

These rock masses are chilled and barren environments, uninhabitable by trees and inhospitable to most other plants. Here and there a hardy shrub gains a foothold, and small clumps of grassy vegetation manage to survive among the shifting rocks. There is little to encourage animal residents, and few are rugged enough to live in these windswept rockslides.

The pika is the exception. It spends its entire life among the rocks, enduring the summer heat and the winter snows in the rarefied atmosphere of the mountaintops. It weighs no more than six ounces and resembles a guinea pig in size and appearance. It is chunky, round, and well-furred, with short, broad ears and an inconspicuous tail. It is related to rabbits and hares, and is known by a variety of other names, including coney and rock rabbit. The pika's body is clothed in tan or grayish fur that blends in well with its surroundings. It is difficult to spot the pika as it moves through the countless tunnels and passages that riddle its pile of rocks. If all is well, a series of short squeaks may be heard and a pika may be seen sitting hunched up, its alert little eyes peeking cautiously over the edge of a rock. It sticks close to the slope, conversing with its neighbors and alert for enemies. At the first sign or sound of danger, it will raise a prolonged alarm with a high-pitched cry and disappear like a flash among the rocks. Two of its most formidable enemies are martens and weasels. Each of these can pursue the pika into the maze of crevices. Sometimes a colony of pikas can confound and finally exhaust a pursuer by coaxing it to chase a number of them at one time as they crisscross dizzily through the length of their underground labyrinth. Much of the pika's day is devoted to the search for food. Its hind legs are only slightly larger than its front legs, so that it dashes smoothly over the rocks. In the midday summer heat it will retire for a siesta to a cool retreat under the rocks.

The pika's burrow is never far from the piles of hay it

harvests for its winter food supply. The pika has developed the unique habit of gathering quantities of food and piling them on flat rocks for curing in the sun. The growing season is short, so that huge quantities of food must be harvested quickly before the first frost destroys the crops. The summer days, except for siesta time, are spent rushing back and forth, bundles of forage in its mouth, from the "hayfields" at the edge of the rocks to suitable drying areas near its home. Most of the forage consists of grasses and herbaceous plants and some seeds and berries, all growing close by, for the pika rarely ventures more than one hundred feet from its rock pile. The bundles of food in its mouth are often larger than the animal itself, but the awkward loads are carried along trails that are familiar and well defined. The hair on the soles of the pika's feet provides a firm footing on the bare and slippery rocks. Before the cold sets in, the cured hay is stored deep in the rockslide. There is plenty of food and no need to hibernate. The little haymaker of the mountains goes about its business despite icy winds and deep snows. Its coat of fur is thick and heavy. Even its ears and feet are well covered. In its subterranean nest, surrounded by its nearby harvest, the pika gets through the winter in good order. By springtime, its stocks of food are largely gone.

The yellow-bellied marmot, a close relative of the woodchuck, will often share the pika's habitat. Though this large, brown, lumbering rodent with yellow underparts has a very wide altitudinal range, it has discovered the advantage of making its burrow among the rocks and boulders near the timberline into which it cannot be easily pursued by such predators as bears, mountain lions, bobcats, red foxes, and short-tailed weasels. Each marmot usually has several burrows, one used as its home and the others as escape routes. It is generally active during the daytime, and its loud, clear whistle is an unmistakable sound of the high mountain peaks and meadows. It often begins its day with a sun bath on the lookout boulder above its burrow, conversing with its neighbors and alertly surveying the terrain for any possible danger. It is similarly on guard while basking in

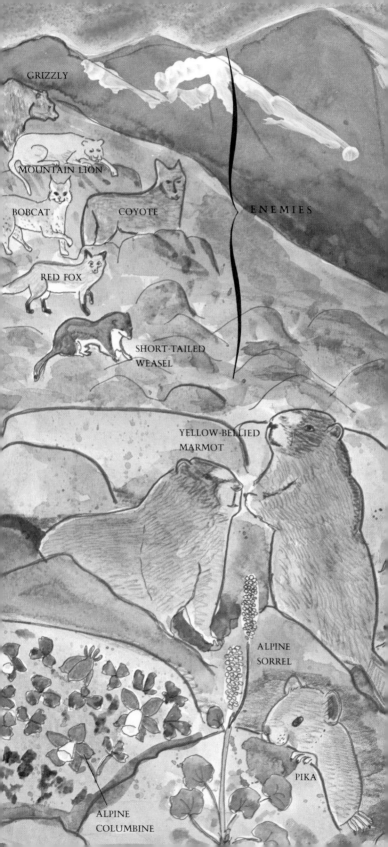

GRIZZLY

MOUNTAIN LION

BOBCAT COYOTE

ENEMIES

RED FOX

SHORT-TAILED
WEASEL

YELLOW-BELLIED
MARMOT

ALPINE
SORREL

PIKA

ALPINE
COLUMBINE

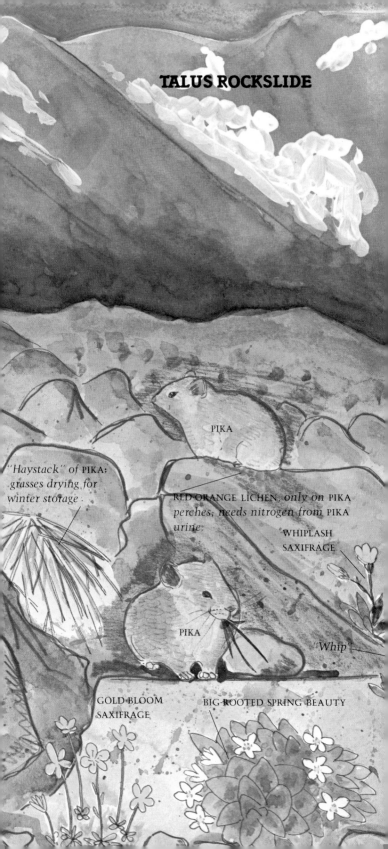

TALUS ROCKSLIDE

PIKA

"Haystack" of PIKA; grasses drying for winter storage.

RED-ORANGE LICHEN, only on PIKA perches; needs nitrogen from PIKA urine.

WHIPLASH SAXIFRAGE

PIKA

"Whip"

GOLD-BLOOM SAXIFRAGE

BIG-ROOTED SPRING BEAUTY

the midday sun. When threatened by an enemy, it will stand on its hind legs "picket-pin" or bowling-pin fashion, whistling high-pitched warning signals at short intervals before fleeing into its burrow. Its size and strength permit it to forage over a wide range; when menaced away from the safety of its burrow, it will threaten an adversary with its long sharp claws. When necessary, it will defend itself courageously.

Unlike the pika, the marmot does not have to spend all day hustling about to harvest a supply of food for winter. It spends the spring and summer months eating large quantities of grass, berries, and other meadow plants, laying on an enormous amount of fat. In the fall, it digs below the frost line, lines its den with soft grasses, and then crawls in for a winter of hibernation, the length of which depends primarily on the elevation. On the higher mountaintops, the period of deep sleep is longer than in the warmer areas of lower regions. Curled up in hibernation in a compact ball, its body temperature and heartbeat are reduced drastically. The store of fat will sustain its reduced bodily functions until warm weather returns, when the marmot will awaken in a hurry to begin the annual search for a mate and the repetition of the fattening process.

With all of the rock fragments moving down the mountainside, some of the smaller particles, especially those resulting from the disintegration of the rock surfaces, collect in the pockets and crevices of the talus slopes. Organic matter is added from the decaying tissue of lichens, sedges, and grasses that grow above the timberline. In time, small patches of soil are ready to support wildflowers hardy enough to live in the high mountains. One of the earliest to bloom in the springtime, as soon as the snow is gone, is the big-rooted spring beauty. This is a perennial plant that grows from a deeply buried carrotlike root, and produces a five-petaled, dark-veined flower, usually white in color. Various species of saxifrage are also tucked among the rocks. Gold-bloom saxifrage, a plant no bigger than two or three inches, has five yellow petals marked with orange spots. Whiplash saxifrage, also yellow-flowered, is named for its runners, which take root and form additional shoots. Like many of the tundra plants, some of the saxifrage flowers are circumpolar plants that moved southward in the path of the continental ice sheets and remained behind in the mountains of North America, Europe, and Asia when the glaciers retreated slowly to the north. The beautiful blue columbine, with its distinctive spurs, has a miniature drooping species that is usually found tucked away between the rocks at these high altitudes. Alpine sorrel grows in the moist, rocky areas. It has round or kidney-shaped fleshy leaves on long stalks and a stem as much as a foot tall that bears small greenish to red flowers.

Barren rock piles are a fairly common sight high in the mountains. Sometimes they are not talus slopes but mine tailings, the debris of man's quest for minerals. The large fragments that make up the talus seem to be firmly fixed in place. In fact, they may remain undisturbed for decades or even centuries. But in the pageant of geology, they provide only a temporary home for plants and animals. Mountain climbers know to skirt these unsteady structures. They will break up and start moving again in a relatively short time, crashing farther down the mountain like a glacier of rock. Weathering and gravity are the lurking and inevitable victors.

TIMBERLINE

Toward the summit of the mountains, the spruce-fir forest becomes less dense and the trees become fewer in number and shorter in size. Living conditions higher up are too severe for trees. The temperatures are low, the winds are desiccating, the growing season is short, the moisture is limited, and the snow cover is minimal. Here and there a single lopsided tree stands out, with no branches on its windward side. Other trees are stunted and twisted, making a last stand against the inhospitable habitat ahead. Gradually, even the most tenacious of the trees must yield, hugging the ground and huddling together in mass resistance to the elements. This is the timberline, the upper limit of tree growth.

The shift from forest to the treeline is not dramatic. It is a gradual transition, with growth often dependent upon minor climatic features. In Rocky Mountain National Park, timberline occurs between 11,000 and 11,500 feet. It is found at lower and lower altitudes toward the north, at 9,000 feet in Montana and at only 6,000 feet in Glacier National Park. Altitude is not the sole determinant. The treeline is very irregular. It is higher up on warm, south-facing slopes and in protected ravines, and lower on the cold north-facing exposures and windswept ridges. Only the most rugged tree can grow in this windy, frigid height. Some Engelmann spruce and subalpine pine survive in modified forms. Other characteristic timberline trees are the bristlecone pine, limber pine, and whitebark pine. Some hardy shrubs keep these trees company. All take on typical features of the environment and create patterns of growth that have inspired such descriptive terms as *elfin forests*, *wind timber*, *banner trees*, *flagging*, and *krummholz*, the latter being the German word for "crooked wood."

Some trees have their own way of coping with these austere conditions. The subalpine fir, true to its name, marches up the mountainside, seeking a shelter from the cold and wind. This may be a depression carved out by glacial ice, or the sheltered side of a cluster of rocks. As it moves up into these higher altitudes the fir tree becomes smaller, but it retains its shape because of the protection of its environment. At the point of its farthest advance, it is a mere miniature of itself, an elfin replica sometimes no more than a foot or two high.

In exposed areas, where living conditions are less favorable, there are isolated patches of so-called wind timber. Here Engelmann spruce grow in stunted mats that are almost at ground level on the windward side and sometimes as much as eight feet tall on the downside slope. The compact growth is an adjustment to wind, frost, and snow. The snow line dictates the height of growth, but the clump keeps advancing in the direction of the wind. The result is a tangled mass so compactly woven that one tree cannot be distinguished from another. The formation takes the shape of a low shed built into the slope, with a roof strong enough to permit "walking on top of a forest."

Here and there in this wind-beaten environment, the "banner" trees stand tall, despite the hardships, like sentinels on mountaintops. These are Engelmann spruce with no branches on the windward side. On that side, wind, hail, frost, and bright sunlight are the victors, and the young buds the vanquished. The branches grow on the leeward side, and they stream out from the trunk like a banner in the breeze. If a branch does take hold in the face of the wind, it can survive only if it curls around the trunk in the opposite direction.

The snow also contributes to the pruning of timberline trees. New shoots that grow above the winter's snow level will not survive. Year after year, these leader shoots are doomed, killed by wind blast and frost if they project above the snow line. But the branches unexposed in winter continue their outward horizontal growth and, as they elongate, become flattened by the weight of ensuing snows. Their height is stable, never rising above the snow line even after decades of growth.

Other trees on the timberline are picturesque spectacles, gnarled and twisted trunks scoured of their bark by the frigid winter winds and whipped by the gale-driven sands of other seasons. Some of them seem to be growing on untenable heights, their roots forced into narrow cracks in the bedrock. Standing in solitude, a "loner" on the highest exposed ridge or peak may be an old limber pine, a stunning example of the survival of the fittest, its limber branches bending with the wind and its smooth trunk, almost at a right angle, majestically defying the elements. Many of its branches may be extra long, twisting like snakes, with tough, flexible twigs that sometimes become tied into knots. The needles of this pine, arranged in bunches of five, are thick and stubby and grow only at the end of the twigs. Its cones when mature are columnar in shape, light brown, and without prickles.

Sharing these exposed and rocky situations, singly or in open groves, are the bristlecone pines, trees with life-spans measured in hundreds, and sometimes in thousands, of years, placing them among the oldest living things on earth. The twigs on this tree are brushlike, densely covered with needles, which accounts for its alternate name, *foxtail pine*. Its needles are also arranged in fives, but are short and marked with small resin spots. Its cones, smaller than those of the limber pine, are deep brown in color with incurved prickles at the tips of the scales. The contorted forms of the bristlecones, with their broken crowns and their massive trunks, tell a tale of centuries of struggle and adaptation to endure the adverse conditions of the forest frontier.

In the more northerly regions of the Rockies, whitebark pine may often be wind-pruned into a stunted, sprawling shrub in the severe climate of the open heights. Like the other two pines, the needles of the whitebark pine are also arranged in fives, but its cones, with triangular-tipped scales, disintegrate at maturity. The grayish luster of its thin bark is the reason for the name *whitebark*. A seedling will often seek the shelter of a boulder or some other natural windbreak and then, as it grows, will bend its trunk and its branches close to the ground, seeking maximum advantage from its favorable location.

Much of the timberline belt consists of extensive unforested areas flooded with bright sunlight. The winters are long and cold. The solidly packed snow stays late and melts very slowly, some of it remaining well into the summer. There is very little rain, and most of the moisture available for plants comes from springtime meltwater. The soil itself at these high elevations is thin and fragile. The growing season is very short, but grasses and sedges sprout and mature in a matter of weeks. Perennials have an advantage, with their woody rootcrowns ready to shoot up as the grip of winter slackens. There are small, protected meadows with brilliant flowers watered by melting snow. Some species of alpine plants work their way down into these meadows. Blooming usually occurs late in summer. Scattered among the rocks or in the crevices of boulders may be clumps of the fragrant alpine lady fern.

The more widely distributed shrubs include dwarf blueberry, a low sprawling plant with pink-and-white, urn-shaped flowers that ripen into delicious bluish-black berries in late summer, and red elderberry with pyramid-shaped clusters of white flowers and orange or red berries. Rock willow, one of the many species of willows found in the Rockies, is a small six-inch shrub at timberline that trails off into a ground-hugging miniature when it reaches into the tundra. All of these shrubs have learned the lesson of wind pressure. It decreases close to the surface of the earth, so that the prostrate, not the upright, plant is more likely to survive in these wind-battered environments.

All of the conditions that make it difficult for plants, work with equal force against the animals. They must contend with the same extremes of wind and cold, as well as with the treacherous terrain and the scarcity of food and water. Some animals, like ground squirrels, deal with these problems by retiring to a burrow during the worst part of the year and living in hibernation on

Tangled grove of BRISTLECONE PINE

GRAY WOLF

WHITEBARK PINE

LIMBER PINE
A "loner"

PARRY PRIMROSE

ALPINE LADY FERN

CLARK'S NUTCRACKER

DWARF BLUEBERRY

WOLVERINE

WHITE-CROWNED SPARROW

TIMBERLINE

(K R U M M H O L Z)
Elfin timber, miniature forest typical of subalpine fir sheltered by rocks.

Wind timber and banner trees typical of ENGLEMANN SPRUCE.

MOUNTAIN GOAT *introduced from Canada in Colorado and Black Hills, South Dakota.*

BIGHORN SHEEP

PARRY CLOVER

HARRIS SPARROW
(Biggest sparrow)

WILSON'S WARBLER

ROCK WILLOW

LINCOLN'S SPARROW

accumulated fat. Many of the other rodents, active throughout the winter, seek shelter in underground nests and runways, protected and warmed by thick layers of snow. Some of the animals take advantage of their mobility and begin to move down into the protected thickets when the assault of winter begins. In the summertime there are animals that wander up from the forests below. Most of the birds are migrants, and even those that spend much of the year near the timberline abandon their summer quarters for the warmth and relative abundance of the lower zones.

One of the animals thoroughly at home in the more northerly regions and well adapted to living on the most inaccessible ledges is the mountain goat. Though actually related to members of the antelope family, the long fur of its shaggy white coat and the beard of the male give it the look of an oversized goat. Both sexes have black horns that are smooth and short and tilt back slightly. They descend into the spruce and fir below the timberline only rarely, grazing on grasses, sedges, and wildflowers in the summer and browsing on exposed twigs in the winter. They are among the most sure-footed of all animals and calmly pick their way where no false step is permitted and no predators can pursue them. Their hooves are well adapted to the steep slopes. They are fitted with non-skid pads and spread apart toward the front for a firm grip on jagged surfaces. They are at ease on the narrowest of ledges.

Bighorn sheep also spend much of their time in these alpine retreats. They are darker in color than those encountered on the desert mountaintops. Quite independently, they too have developed the type of hoof that makes travel possible over these hazardous areas. Both of these animals are well clothed. The mountain goat has a woolly undercoat beneath his shaggy exterior, and even the straight hair of the bighorn sheep is an adequate shield against the cold. Both are wary animals, remarkable in their ability to survive in one of the earth's most inhospitable environments.

There is no shortage of predators. Owls, hawks, foxes, and weasels depend on the rodent population. There is a steady supply of ground squirrels, gophers, and marmots. Mountain voles defend themselves simply by multiplying. At times their prolific breeding creates the danger of a self-destructive population explosion. This is when the carnivores have a feast and themselves increase in numbers. But at a certain point the vole population collapses, and their predators are put under a strain.

At the top of the food chain are bears, bobcats, mountain lions, wolverines, and gray wolves. The wolverine, an inhabitant of the more northerly regions, is an aggressive enemy, strong enough to pull down an animal three times its own size. It is a real glutton and its reputation for eating immense quantities of food has earned for it the Latin name for glutton. Another northern predator is the gray wolf, which will feed on small rodents when alone, and on big game when hunting in packs. They will not attack a herd, but they follow patiently until a healthy animal can be isolated or until a sick or lame animal falls behind, and then surround their victim for the kill.

There is something impressive about the timberline. It is not an ecological zone. It is a transition area, or ecotone, between the tundra and the forest below. All ecotones are points of contact between contiguous communities, where the intermingling of species from two different neighborhoods adds a special richness and variety to the landscape. But where the trees end there is the unscarred spectacle of snow-capped peaks and ranges standing stubbornly against the bright-blue sky. This is a place of wonder—a place called timberline.

ALPINE TUNDRA

Just beyond the twisted, wind-beaten, and ground-hugging growth of the timberline is the high-mountain environment called the tundra. Here climatic conditions due to altitude correspond to those due to latitude in frigid zones. But the tundra is not a uniform environment. Spread out in all directions are patches of well-established sedges, scattered tussocks of grass, rock-strewn meadows, fell fields of boulders, meltwater bogs, and patches of soil disturbed by weather or burrowing animals. In southern Colorado, these conditions begin at an altitude of 11,500 feet; in Montana, they start at about 8,000. Since this zone is on top of the mountains, it is called alpine tundra, but living conditions are fairly similar to those in the Arctic tundra.

This is no haven of comfort. Fierce winds, extreme temperatures, unstable soil, and a short growing season are constant factors. Much of the region is free from snow for most of the year. The high winds, which increase with altitude, drive the snow from the exposed places into protected depressions where accumulated drifts are trapped for a good part of the year. After a long, freezing winter, the parching summer sun may raise the surface temperature well above 100 degrees. But at high altitudes the rarefied atmosphere dissipates the earth's heat quickly, bringing the nighttime temperature back to zero. Near the snowbank sites, a steady supply of water is generated by the annual thaw. In other areas, summer thunder showers with their hail and sleet provide a large supply of moisture, but this is only transient. The slopes are steep and the soil is porous, so that drainage is rapid. The winds accelerate evaporation and dehydrate the plants. On some slopes, large concentrations of snow come hurtling down as avalanches, which can destroy a mass of vegetation in moments. Rockslides, the result of the weathering process, can uproot or bury plants that have been established for decades. The winter is long, spring is merely a flash, and summer is half the season it is on the plains. Yet in this short span the barren fields are ablaze with flowering color and astir with animal life.

The dominant vegetation of these tundra regions is low-growing wiry sedges and grasses. Flowering plants become established around the outcrops of rocks, in gravelly sections where weathering has broken up the turf and where the soil has been disturbed by other causes. Annual plants are a rarity. It is too much to expect a plant to rush through a complete life cycle of germination, growth, flowering, and seed production in the brief summer days that run the gamut of extreme weather changes. Plants that do not have a steady supply of water from melting snow live in desertlike conditions. The shortness of the growing season imposes a perennial existence on most of these plants. Even the perennials have a hard time, and survival depends on a multitude of special adaptations.

The most fascinating and successful of the tundra flowers are the so-called cushion plants. All of these plants have learned to cope with desiccating winds, low temperatures, and the limited supply of moisture. These are slow-growing plants, lying dormant most of the year. In any one growing season, a typical seedling may put out one or two minute leaves. It may take a decade or two for them to mature, and a plant only six inches in diameter may be more than one hundred years old. These plants rarely grow more than a few inches tall. Most of their energy goes into the production of large root systems that anchor the plants on the gravelly slopes. Their taproots go down into the soil two feet or more to locate water and to absorb and store it rapidly whenever it becomes available. The cushion shape is vital. The rounded, short-stemmed form presents the least vulnerable surface to the abrasive and gouging action of wind, snow, and hail. The slightest depression in the ground or the smallest collection of pebbles provides protection for these ground-hugging miniatures. Heat absorbed during the daytime is trapped inside the closely matted foliage and conserved for nighttime warmth. Each little cushion is a mixture of new and past growth. Old organic matter consisting of dead leaves, old seed capsules, and decaying stalks provides humus and adds to the compactness of the cushion and its ability to sponge up water and prevent evaporation. Insects, benumbed by the cold, find refuge in the warm interiors of the cushions. They repay those of their hosts that require it by performing the vital function of pollination, thus ensuring the perpetuation of the species.

There are other adaptations that contribute to the survival of these tundra plants. Many of their blossoms are brilliantly colored. The deeper the shades, the greater the ability to absorb heat. Intense colors may even have a special fascination for certain insects. Some flowers are pale or white, but most of the plants have dark-green heat-absorbing leaves. The leaves may have a number of other special common characteristics. Some are smooth with a waxy finish that resists evaporation. Others are woolly, with a dense, hairy fuzz that tempers the chilling effect of the wind. There are fat and succulent leaves well equipped to store water. All of the leaves are small, many have inrolled margins, and most have evenly continuous edges. These are features designed to preserve a plant's moisture. The combination of these various adaptations enables many of these plants to survive in this region.

The cushion plants are the most exciting. When one of these plants finally comes into its first bloom after years of patient effort, it may celebrate its victory with a display of hundreds of brilliant blossoms over a surface no larger than a dinner plate. In the gravelly meadows are the yellow-eyed, sky blue, or white flowers of the alpine forget-me-not; the stemless pink blossoms of the moss campion, or moss pink; the pale blue or white five-petaled flowers of alpine phlox; and the rose-colored or purple flowers of the dwarf clover. Many of the plants in the Rocky Mountain tundra are known as circumpolar species, related to arctic varieties that owe their wide distribution to the last Ice Age. These plants are presumed to have come south in the path of the great glaciers and to have been driven down onto the foothills and plains. As the ice sheets retreated, some of these plants remained behind but climbed up the slopes of the mountains as the weather became more moderate. Now they have found a place in the high elevations of the temperate zone, where conditions are comparable to those at sea level in the Arctic. The large number of similar plant species in each of these regions suggests the possibility that the Arctic and alpine tundra were once a continuous community.

The high meadows and well-drained slopes sparkle with plenty of other color. Hundreds of rydbergia, or alpine sunflowers, with their large yellow blossoms and gray, woolly leaves, may cover a rocky ridge, all facing in the direction of the rising sun. This is why they are sometimes called compass flowers. One of the more common plants is alpine avens, whose five-petaled yellow flowers bloom in profusion early in the summer, and whose leaves turn dark red with the first signs of autumn. Many of the tundra flowers are shorter-stemmed and fuzzier varieties of species living at lower elevations. In the normal evolutionary process some plants develop characteristics more appropriate to colder and drier environments. In time, a single genus might move through several mountain zones, leaving a representative in each of them. Phlox, pinks, and sunflowers are to be found all the way from the foothills up to the highest elevations. The alpine harebell, with a single flower on each stem, is the diminutive cousin of the many-flowered plant that reaches two feet or more at lower altitudes.

In some areas, voles, gophers, and other small burrowing animals dig under and destroy part of the tundra vegetation. These patches have the appearance of cultivated gardens. The sky pilot, with purple, funnel-shaped flowers and contrasting orange centers, is one of the plants that take root in these mounds of dirt. Where the meadows are moist or wet, the plants may be members of the family of succulents. One of these, with the usual fleshy leaves, is rose-crown, or queen's-

ALPINE TUNDRA

BLACK
ROSY FINCH

GRAY-CROWNED
ROSY FINCH

BROWN-CAPPED
ROSY FINCH

WESTERN YELLOW
PAINTBRUSH

RYDBERGIA
(ALPINE SUNFLOWER,
OLD MAN OF THE
MOUNTAIN)

KING'S CROWN

FAIRY
PRIMROS

ALP LILY

ALPINE
HAREBELL

A L P I N E   M E A D O W   P L A N T S

ALPINE
FORGET-
ME-
NOT

ALPINE AVENS

LICHENS

WHITE-TAILED PTARMIGAN

Male

Female

WATER PIPIT

MOUNTAIN DRYADS

SMINTHEUS

QUEEN'S CROWN

SKY PILOT

SNOW LILY

OCK-ARDEN LANTS

SNOW ACCUMULATION AREA PLANTS

MOSS CAMPION

SNOW BUTTERCUPS

crown. It grows in dense clusters and has small rose-colored flowers that look like clover. King's-crown, somewhat similar in appearance but with darker flowers, prefers the less marshy areas. In the dry boulder fields, between the rocks or in the crevices, grows the small fairy primrose with yellow-centered, purple flowers. There is also alp lily, a small lily with a single bloom on each stem. The six segments of its flower are whitish with purplish veins. On the exposed ridges and gravel slopes are two plants that often grow together. Alpine willow, a dwarf version of a shrub at lower elevations, forms large, ground-hugging clumps. The mountain dryad has a similar prostrate habit. Its shrubby mats often spread over rocks and boulders. It has a large white flower, usually with eight petals, at the end of a short stalk. Its seeds are produced in a round, plumy head.

The snowdrift sites have a totally different community of flowers. Snow is often a protection, not an enemy, of plants. The surface of a snowbank may be well below zero, but the bank itself acts as a protector against the wind and as an insulator that keeps the temperature of the soil at a uniform level and warmer than the outside. Despite the generally adverse conditions, some delicate perennial plants do not freeze under the winter snow. They push above the edge of the drifts as soon as the melting begins in the spring. Alpine or snow buttercups pop up almost overnight like magic and develop quickly to produce leaves and bright yellow flowers within the limited growing season. If the snowbank fails to melt one year, the buttercups will conserve their energy and wait until the next. The snow lily, or dogtooth violet, with its nodding yellow flowers, blooms at the lower elevations early in the spring, as soon as the snow recedes, and continues to bloom into midsummer, at higher and higher elevations, as it pursues the melting snowline up the mountainside. But not all snow is helpful to the flowers. Some on the north-facing slopes does not melt until too late in the summer for any plants to grow, and some of the snow cover is too thin to provide adequate protection.

On the heights above the tundra vegetation, polished rocks are covered with a soft, crustlike growth of lichens—the Lilliputians of the plant world. These minute plants are able to live on bare rock, on the bark of trees, on the walls of canyons, on dead timber, on wet soil, and in the cold ground of the Arctic. Wherever they grow, they play a role, along with wind, rain, heat, and frost, in the leisurely process of soil building. But in the high mountain peaks, lichens enter into the ecological process of succession at its most primeval stage. Here, among the exposed boulders of granite and sandstone where nutrients are at a minimum, these ingenious nonflowering plants are crumbling the rocks into minute particles on their way to becoming part of the soil.

Lichens are a composite organism made up of a fungus and an alga, each dependent upon the other in a mutually beneficial union. The fungus provides body and protection for both by anchoring itself to a rock. The alga, through the process of photosynthesis, is the food-producing unit for itself and the fungus. Acids secreted by the bodies of lichens help to decompose and dissolve the rocks. Tiny bits of sand are trapped by the plants. In time the lichens die and their decayed organic remains are added as humus to the grains of sand they have pried loose from the mountains. Soil building has begun on the rocky tundra slopes. The lichens that spread out on the rocks in a delicate lacework of green, gray, yellow, orange, or black are the true pioneers of the plant kingdom.

There are not many permanent animal residents of the tundra. Mountain goats and bighorn sheep graze and browse on the vegetation but move down into the lower elevations in winter. Mule deer and elk occasionally venture up the slopes in the summertime. Those that remain all year are the various rodents that live in burrows, and the pikas and marmots that have their dens among the large boulders. Three birds are closely associated with the alpine terrain—white-tailed ptarmigans, rosy finches, and water pipits. The mottled summer plumage of the white-tailed ptarmigan provides good protective coloration. In the winter its plumage changes to pure white, except for its bill. The terminal buds of the alpine willow make up its staple winter food. It burrows into the snow for protection but at times moves down to lower levels. The rosy finches and water pipits breed in the high mountains and spend much of their time walking in the snowfields searching for insects and seeds. In late fall and winter, they migrate to warmer climates.

The tundra is a breathless place for humans. One's breathing rate must increase to compensate for the reduction in oxygen. Blood chemistry changes. The slightest exertion produces a tired feeling. Not many make the climb. The higher the ascent, the lower the temperature. The weather is unreliable, usually at one extreme or another. It is a place for a short visit, best left to the plant and animal species that have taken a millennium to adjust to its severity.

SONORAN DESERT

Deserts have certain common characteristics. Water is scarce, average temperature is relatively high, and humidity is low. Basically, any region with less than ten inches of rainfall, unevenly distributed throughout the year, qualifies as a true desert; some deserts receive as little as three inches. Even the rain that does fall is not always available to the plants and animals. Some of it is lost through rapid evaporation, aided in many regions by desiccating winds. Much of the rainfall comes in sudden cloudbursts. Part is quickly drained off in porous areas, and some is wasted as runoff on sunbaked surfaces. The soil itself is not secure. Rain and wind combine into a powerful agent of erosion, causing the shifting of desert sands and the formation of broad washes and gullies. There are extremes between daytime and nighttime temperatures, and substantial seasonal variations. Sand and dust storms move across the land slicing into the vegetation and driving wildlife to shelter.

Despite all this, a wide variety of life-forms has adjusted to these desert conditions, revealing amazing resourcefulness in meeting the challenges of extreme heat and water scarcity. Some of the adaptations for survival are fascinating. There are desert animals that can go through an entire lifetime without a drink of water. There are plant seeds that can wait in the ground for years until the return of the right conditions for germination. There are animals that virtually "swim" under the loose desert sands.

The deserts of the West are not a single unbroken environment. There are three major regions, each with its own combination of climate, rainfall, and soil. The largest desert is the Great Basin, which includes land between the Rockies and the mountains of California, much of it lying in Nevada and Utah. In southern California, with portions extending into Utah and Arizona, is the comparatively small Mohave Desert. To the south and east of the Mohave is the Sonoran Desert, extending from Mexico into southern California and western Arizona. California's part of the Sonoran Desert is sometimes classified separately as the Colorado Desert, named after one of its boundaries. The Sonoran is about 120,000 square miles, consisting mainly of sandy or rocky plains dotted with various detached mountain ridges. The silt from these mountains spreads down into the plains in broad, gentle slopes called "bajadas." The Sonoran has both summer and winter rain. This and its mild winter weather contribute to a more abundant and varied vegetation than is found in the other deserts. Because of the number of small trees, the mixture of evergreen and deciduous shrubs, the diverse species of succulents, and the rich displays of wildflowers, the Sonoran is often considered the most magnificent of American deserts. Plants as well as animals fill their respective niches, each adjusting to the rigors of heat and drought in its own special way.

Plants that have adapted to an environment in which water is scarce, as in low-lying deserts or on wind-swept mountaintops, are called *xerophytes*—a term derived from the Greek words meaning "dry plants." Survival is generally achieved through structural or behavioral adaptations, or both. There are plants that patiently escape the conditions of drought, those that resort to methods of evasion, and those that simply stand and resist.

Those that escape the conditions of drought are the flowering annuals. Their technique is a compressed life cycle. Their seeds remain dormant, sometimes for more than a year, waiting out the dry periods. When the right combination of sunlight, moisture, and temperature arrives, a burst of growth begins. With feverish activity, the seeds germinate, the plants grow, and the flowers blossom, all in a matter of a few weeks. Then the plants go to seed, and another waiting period begins. Many of these seeds will not germinate until their outer covering is dissolved by the required amount of moisture.

These short-lived annual wildflowers, sometimes called "ephemerals," are usually at their peak in April in a sparkling floral spectacle. Because of the many variables involved in producing the right conditions for growth, an outstanding display does not occur every year. Annual wildflowers also bloom soon after the summer rains, and these are usually at their best in August. The seeds of these late-bloomers will also remain buried in the sand until their miraculous rebirth at an appropriate later season.

Evaders are those plants that have the problem not only of securing a supply of water but of preventing its loss by evaporation from their leaves. The ocotillo does this by simply shedding its leaves as the soil dries out. Sometimes called the coachwhip, the ocotillo plant is a cluster of slender, thorny stems, often more than ten feet high, each of which in the spring bears a brilliant scarlet flower cluster at its tip. After a rain, small leaves crop out around each thorn, but these fall off during the dry season. It can grow leaves several times a year following a rain and slowly go dormant thereafter.

Three tree-sized plants, each with its own technique for preserving moisture, produce bean pods that are a favorite food of many desert animals. The mesquite can be a spreading tree in the more favorable drainage situations, or a shrub in the drier zones. To reach a dependable supply of water, its root will descend to a depth of forty feet. This strong taproot together with the lateral roots anchor the mesquite in its sandy environment and protect it against blasting desert winds. Its thorny wood, hard and heavy, is a popular firewood, and its special aroma is typical of the desert. To avoid excessive evaporation, the leaves of the mesquite curl up lengthwise, thus exposing a smaller surface to the sunlight. There are two kinds of mesquite, the "honeybean" and the "screwbean." The spiral shape of the screwbean pod distinguishes it from the honeybean. The paloverde, often growing close to the mesquite, puts out small leaves in times of rain. The plant drops these leaves during dry spells and depends on its green trunk and stems to carry on photosynthesis. (*Palo verde* in Span-

SONORAN DESERT

MESQUITE

PHAINOPEPLA *on* MISTLETOE

VERDIN

OCOTILLO

VERMILLION
FLYCATCHER

MESQUITE
BEANS

DESERT POPPY

PARRY PENSTE

KIT FOX

DESERT
MARIGOLD

NIGHT-BLOOMING
CEREUS

SIDEWINDER

KANGAROO RAT

ish means "green stick.") In the spring, the tree is a spectacle in yellow as thousands of tiny flowers cover its wide-spreading branches. Both the foothill paloverde and the blue paloverde have been designated as Arizona's state tree. A third plant, and the hardiest of them all, is the ironwood. It loses its leaves only in very dry periods, relying generally on the tough, shiny surfaces of the leaves to reduce the loss of water. Green clumps of desert mistletoe are often seen in the upper branches of these three trees. This plant is a partial parasite, for though it can manufacture its own food, it must depend on its host for water and minerals. The berries of the mistletoe are an important food for birds.

The most remarkable xerophytes of the desert community are the cacti—the so-called resisters. These plants have substituted spines for leaves, through which most plants lose their moisture, and transferred to their green stems the work of producing food. The stems of many species have become more and more barrel-like in order to minimize the proportion of surface area to total volume. The coating of the stems is thick and shiny. This not only saves water but protects the plants against the cutting action of sand storms. They have wide-spread lateral root systems that can absorb great quantities of moisture during the brief periods of rain. Their roots and stems are filled with water-storing tissue. They have diversified into many species in an amazing variety of shapes and sizes. They provide shelter for desert creatures. Many of their fruits are edible and flavorful. In bloom they brighten the barren desert with a brilliant profusion of color.

The structure of many of the cacti takes the form of jointed stems. Prickly pears are everywhere. These are the cacti with flat, jointed pads, covered with sharp spines and dense bristles. They grow sprawled over the sand, in clumps as shrubs, and sometimes as small trees. The flat sides of their joints often have an east-west orientation, exposing only the thin edges of the pads to the midday sun. Some animals devour the stems. Many make the fruit a reliable part of their diet. Many of the cacti are cylinder-jointed. The main branches of the

Christmas cactus consist of such cylindrical joints. The bright-red, fleshy fruit remains on the plant into the winter months when the desert is drab—hence the name *Christmas*. The adult plants of buckhorn cacti are small trees with outer branches that look like antlers. Pencil cacti form dense thickets. Their stems are finger-size, and their fleshy green fruits remain attached to the plant for about a year.

Two species of the cylinder-type cactus are referred to as "jumping" cactus, because their sections are loosely joined and seem to fly off the plant with the slightest jar, and because just brushing against the plant will transfer some of the needles to a person's clothing or body. One of the jumping plants is chain-fruit cholla. This treelike cholla may grow up to twelve feet in height, with long chains of drooping fruit that hang on year after year. The seeds of this plant are rarely fertile, so the plant reproduces vegetatively from joints, which send out roots after they fall to the ground. The other jumping cactus is the smaller teddy-bear cholla. It has a thick-set armor of whitish spines that glistens in the sunlight with a soft, thistly, and deceptively harmless effect. One of the most beautiful flowers in the cactus world is produced by the night-blooming cereus. Its many-pointed, starlike blossoms open just after nightfall and close up early the next morning. Unlike other cacti, these plants store their water underground in a thick, turnip-shaped taproot that may weigh fifty or more pounds. The ribbed stems of this plant look like a series of dried sticks.

The other group of cacti are ribbed and unjointed, in the shape of either a cylinder or a globe, or in modifications of these forms. The largest of these plants are the barrel cacti, some as tall as six feet and several feet in circumference. Barrel cacti have long been credited with the ability to store free water. Actually, the plant's moisture is contained in a sticky pulp which is definitely not palatable to humans. It is sometimes consumed by desert bighorn sheep, who break open the cacti with their strong horns. Hedgehog cacti are smaller in size, usually four to twelve inches. Some are

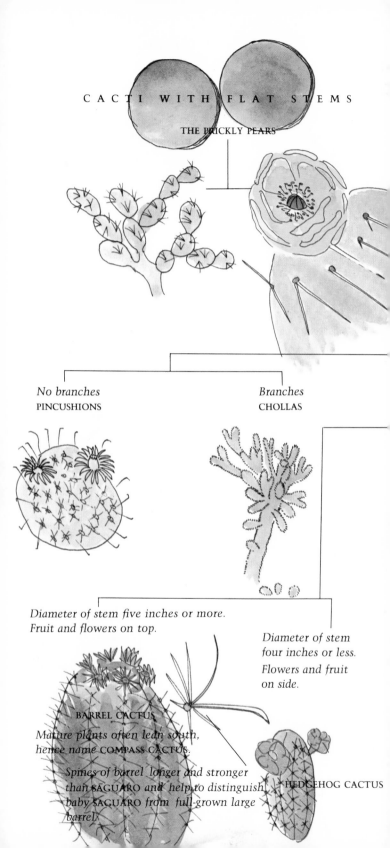

CACTI WITH FLAT STEMS

THE PRICKLY PEARS

No branches
PINCUSHIONS

Branches
CHOLLAS

Diameter of stem five inches or more.
Fruit and flowers on top.

Diameter of stem
four inches or less.
Flowers and fruit
on side.

BARREL CACTUS

Mature plants often lean south,
hence name COMPASS CACTUS.

Spines of barrel longer and stronger
than SAGUARO *and help to distinguish*
baby SAGUARO *from full-grown large*
barrel.

HEDGEHOG CACTUS

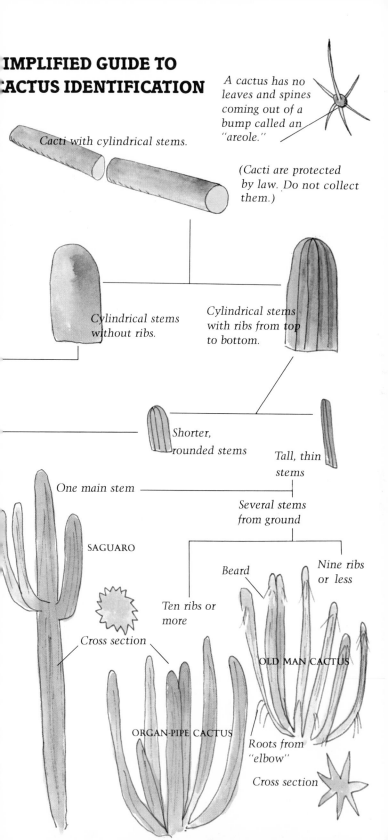

SIMPLIFIED GUIDE TO CACTUS IDENTIFICATION

A cactus has no leaves and spines coming out of a bump called an "areole."

Cacti with cylindrical stems.

(Cacti are protected by law. Do not collect them.)

Cylindrical stems without ribs.

Cylindrical stems with ribs from top to bottom.

Shorter, rounded stems

Tall, thin stems

One main stem

Several stems from ground

SAGUARO

Cross section

Ten ribs or more

Beard

Nine ribs or less

OLD MAN CACTUS

ORGAN-PIPE CACTUS

Roots from "elbow"

Cross section

solitary, while others form tightly bound clusters. The smallest of the unjointed cacti are the pincushions, most no more than one to four inches tall. Some are solitary, some branch sparingly, and others form dense clumps.

Standing majestically tall above all the desert vegetation is the saguaro. It begins life precariously. Hazardous environmental factors account for a high mortality. During the first two years of its life, it is a favorite food of rodents. It needs the protective shade of a "nurse" plant to survive as a seedling. Growth is painfully slow. After ten years it is only four inches high; at fifteen it may have reached a foot; and at sixty or seventy, when it is fifteen feet tall, it begins to send out its first branches high above the ground. At maturity it is a massive plant, reaching a height of thirty-five feet or more, with a life-span of one hundred fifty years.

The preferred habitat of the saguaro is a rocky slope or the gravelly soil of an upper bajada where it can establish an anchor against the desert wind. The immense weight of the top-heavy column is supported by a strong taproot and by a network of outstretched, thickly interwoven lateral roots with a radius of sixty feet. The pulpy interior of the column and of the branching arms is held upright by a cylindrical framework of vertical wooden rods. Here the plant can store a ton of water when it rains, enough to carry it through months and even years of drought. The outer covering of the plant expands in order to absorb the supply of water. In the dry season it contracts so that the ribs become more prominent. The whole system is an ingenious architectural structure functioning as an adjustable storage tank.

The saguaro is the center of considerable animal activity. Its flowers open at night and close the following afternoon. White-winged doves, long-nosed bats, and many insects feed on the nectar of the large, waxy-white blossoms. The juicy red pulp of the fruits, each containing about two thousand black seeds, is a favorite food of desert birds. The Gila woodpecker, zebra-backed with red headpatch, and the gilded flicker, gray-faced with yellow wings and brown crown, dig deep nest cavities in the soft stems. To heal these wounds, the plant produces a hard callous tissue and the opening takes on the shape of a shoe. Abandoned shoes are comfortably cool and do not lack an occupant for long. Elf owls and screech owls, flycatchers, purple martins, and other birds move in promptly. Snakes, rodents, bats, and insects may also use the shoes for shelter and protection. The fork of the upper branches is ideal for the nests of the red-tailed hawk. Pack rats live at the base of the plant.

The animals of the desert must also stay within viable limits of temperature and minimum water requirements. Unlike plants, however, animals are not rooted to a spot. A high percentage avoid heat by staying underground. Many are nocturnal, searching for food only in the cool of night. Even those abroad in daylight spend much of their time in the shelter of vegetation, moving about only for food or to avoid a predator. Some hibernate or remain dormant when extremes of heat or cold become a problem. Many of the smaller animals obtain much of their water from succulent foods. Larger animals must forage over wide wastelands in search of waterholes.

The kangaroo rat has several unusual water-conserving adaptations, but the most extraordinary one is an internal system designed to manufacture water from its diet of dry seeds. Its body is solidly built and covered with silky fur. Its head is almost as large as its body, with big black eyes for nighttime vision. It is not in fact a rat, since it is more closely related to the pocket gopher, with whom it shares fur-lined cheek pouches. The *kangaroo* part of its name refers to its large, strong hind feet, which are used to propel the animals in flying leaps, especially when alarmed. The small, weak forelimbs are used mainly for stuffing seeds and stems into its pouches. Much of the food is stored in its den, to which there are generally a number of different entrances. The hot summer days are spent in this humid refuge. Foraging for food begins just before dark. So long as there is food, there is water made by extracting hydrogen and oxygen from the carbohydrates in the food. This special talent for chemistry gives the kangaroo rat a wider range of habitat than that available to animals dependent on an outside supply of water.

The reptiles have problems with cold weather, especially in winter, when temperatures can drop precipitously during the night. The chuckwalla, a large lizard with a skin-covering that hangs loosely about its body like an outsized overcoat, spends the day sunning itself. At night it seeks refuge under a ledge or in the crevice of a rock, and during the winter months sleeps in an underground burrow. It has the facility of inflating its body so that it cannot be dislodged from its tight hiding place by its enemies. The beautiful crested lizard, whose crest is a row of enlarged scales along its back, lives in the burrows of other animals during the warm months and hibernates underground for the rest of the year. The Gila monster, the only American poisonous lizard, is heavy-bodied and sluggish. Its upper body is covered with pink-and-black beadlike scales. During the hot summer days it falls into a deep stupor. It spends the winter in a moist burrow in a state of hibernation, consuming the fat stored in its thick tail. The desert tortoise, a grayish-brown vegetarian, digs a solitary burrow for use during summer nights and joins a communal burrow for winter hibernation.

Some of the reptiles have special adaptations for moving through the shifting desert sands. The horned lizard, which like the kangaroo rat can manufacture water for itself, is equipped with special nose valves designed to keep its nostrils clear while it digs in the sand. The sand lizard can "swim" beneath the surface of the sand, thanks to its protective eyelids and the extra traction provided by its broad, scaly toes. The eyes of the sidewinder, or horned rattlesnake, also have protective adaptations. It moves up and over the mounds of loose sand in its habitat in a series of looping, angular body motions.

Some of the larger animals range over wide areas. Mule deer spend their summers in higher elevations and frequent the lowlands only during the winter. A few desert bighorn sheep have survived by spending most of their time on the high crags of the desolate mountains. The javelina, or peccary, must move about, since it is dependent upon a steady supply of water. The nocturnal coyote is everywhere, regulating the rodent and rabbit population by hunting them down. It too holes up in the burrow of some other animal during the hotter hours of the day. During prolonged periods of drought, the coyote becomes a digger of wells, locating water for himself and many other desert creatures.

At times the desert may appear to be a lifeless wasteland. But centuries of evolution have produced a complex and lively community of plants and animals that has solved in one way or another the twin problems of excessive heat and water scarcity.

MOHAVE DESERT

The upland region that lies between the Great Basin Desert to the north and the Sonoran Desert to the south, covering the southern part of Nevada and part of southeastern California, is the Mohave Desert. Elevations in this desert range from three to five thousand feet, except in Death Valley, where hundreds of square miles lie below sea level. Clouds moving from the Pacific on the way to the Mohave lose much of their moisture over the mountains before they reach the desert. The scanty rainfall comes mostly in the winter, although there are occasional summer cloudbursts. The winter brings light snow in some areas. The structure of the mountains produces low-lying basins with north-south orientations that have served as passageways for the migration of species from the Great Basin and the Sonoran Desert. Only a fraction of the plants and animals of the Mohave Desert are strictly confined to that region.

The desert is at its best in the spring following a wet winter. When there is the right balance of adequate winter rains and warm spring temperatures, even the rocky hillsides are carpeted with brilliant patches of color. April and May are the best months to observe wildflower displays, although the season may begin earlier in the lower elevations and continue later at high elevations. If rainfall is scarce, there will be few flowers. But if the proper conditions prevail, germination and flowering are completed without a moment lost before the moisture crucial to seed production vanishes in the heat of the summer.

For many people, the Joshua tree symbolizes the Mohave Desert. It is a spectacular tree, found nowhere else in the world. This tree, actually a giant yucca, attains a height of forty feet, but its distinguishing feature is the way its branches twist and turn in all directions.

There is a legend that the Joshua tree was given its name by the Mormons of California on their way to join the main congregation in Utah. In the shape of the branches and the tufts of shaggy, bayonetlike leaves, these travelers saw the bearded patriach of the Old Testament standing with arms uplifted to the sky, urging them on across the burning desert to the Promised Land. In the sandy soil of the desert, these plants grow slowly, taking decades and decades to achieve a full complement of angular branches. In the dry, rocky areas few ever achieve full height. The tips of these branches have a burst of stiff, dagger-like leaves. Each year some of these leaves die, turn downward, and become part of the shaggy matting over the bark. A mature Joshua tree bears leathery, cream-white blossoms in clusters eight to fourteen inches long. The best displays are in the springtime, but the Joshua tree does not bloom every year.

The Joshua tree is often confused with the Mohave yucca, another large member of the yucca family. While they can be found growing together, the Mohave yucca is more common at lower elevations. The leaves of the Joshua tree are about ten inches long and have very fine teeth along their margins. The much longer leaves of the Mohave yucca are easily distinguished by the abundance of light-colored fibers along their edges. Another yucca of the upper desert is the soaptree yucca, often called Spanish bayonet because of the rosette of long, narrow, pale green leaves that taper to a needlelike point. Its other common name is Our Lord's candle—for the tall stems that carry spectacular clusters of the glistening white, fleshy flowers.

There is an inexplicable life-giving arrangement between the Joshua tree and the yucca moth, an interdependence known as *mutualism*. Neither species can survive without the other. The flowers of the yucca can be pollinated only by this moth, and the eggs of this moth must feed only on the seeds of the yucca. The female moth first collects pollen with her specially designed mouth parts and locates a yucca blossom suitable for egg laying. She carefully rubs the sticky pollen

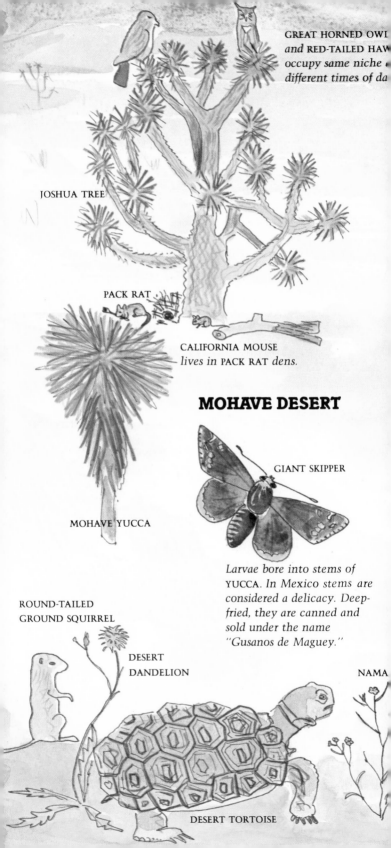

GREAT HORNED OWL *and* **RED-TAILED HAWK** *occupy same niche at different times of day*

JOSHUA TREE

PACK RAT

CALIFORNIA MOUSE *lives in* PACK RAT *dens.*

MOHAVE DESERT

GIANT SKIPPER

MOHAVE YUCCA

Larvae bore into stems of YUCCA. *In Mexico stems are considered a delicacy. Deep-fried, they are canned and sold under the name "Gusanos de Maguey."*

ROUND-TAILED GROUND SQUIRREL

DESERT DANDELION

NAMA

DESERT TORTOISE

JOSHUA TREE

NEST OF
SCOTT'S ORIOLE

LADDER-BACKED
WOODPECKER

SCOTT'S
ORIOLE

CREOSOTE BUSH

OUR-LORD'S-
CANDLE,
a yucca

ANTELOPE
GROUND
SQUIRREL

NIGHT SNAKE

GHT LIZARD

BANDED GECKO

LEAF-NOSED SNAKE

over the flower, and then moves down into the flower, puncturing the ovary and inserting her eggs one by one. This assures a supply of seeds, some of which are eaten by the larvae of the moth while plenty of others mature naturally and are blown over the floor of the desert. In time, the larvae will gnaw their way out of the oval seedpod, drop to the ground, and spin their cocoons. When spring rains assure new yucca blossoms, the moth will emerge, and the cycle will begin again. The timing is always right: the moths appear just when the yucca blooms. Most other yuccas are pollinated in the same way—a process that guarantees food for the young of the moth and seeds for the continuance of the yuccas. No other method of pollination will work, for the yucca pollen is too heavy and sticky to be scattered by the wind or handled by other insects.

As in the case of the saguaro cactus in the Sonoran Desert, the Joshua tree is the center of considerable animal activity. As the tallest tree in the surrounding desert, its upper branches provide an ideal perch for hawks. Fastened to the daggerlike leaves of the tree may be a nest, constructed mainly of yucca fibers, of Scott's oriole. The male, in his vivid black and yellow colors, may often be seen sitting on one of the upper branches. The holes in the trunk drilled by ladder-back woodpeckers become homesites for elf owls. The blossoms, fruits, and seeds, as well as the insects around the trunks and leaves, are food not only for birds but for rats and squirrels. Night lizards hide during the daytime under dead limbs or dead tree trunks. These small lizards feed on the termites in the rotting wood and on flies, beetles, ants, and other small insects in the vicinity.

In much of the Mohave Desert the Joshua trees stand high above endless miles of shrubs. Many of these shrubs are to be found in the other deserts as well. Unlike the cacti, these plants do not store water in fleshy stems but resort to different strategies for hoarding moisture and living through lengthy dry spells. Common to all three of the deserts and covering vast flats is the creosote bush. Each of these plants ensures its own supply of water, whenever sufficient moisture is not available, by eliminating competition around it. It has been suggested that this is accomplished by releasing a toxic substance into the soil, killing seedlings and nearby competitors and causing a wide spacing among the plants. Each plant spreads its roots close to the surface to capture quickly any moisture before it evaporates, and since the plants are fairly evenly spaced, each receives about the same amount of watering. In areas where the subsoil is loose, the roots of the creosote bush penetrate quite deeply. The plant itself has a strong resinous scent. The heavy, varnishlike coating on the small leaves helps to conserve moisture, and their oily appearance explains why the shrub is sometimes called greasewood. Except during periods of severe drought, the bush remains evergreen; even when dried to a lifeless brown, it will quickly return to its deep green color soon after a rainfall.

A shrubby tree that grows over most of the desert regions is the spiny plant called cat's-claw. The smallness of its leaves prevents excessive transpiration. Round clusters of yellow flowers are produced in the spring. Another member of the shrub community is the smoke tree, a deciduous plant that loses its leaves early in the summer. Mormon tea, an unusual shrub related to the conifers, has like the cacti surrendered its leaves in order to prevent loss of moisture. Its jointed yellow branches, covered with small scales, perform the food-making function of leaves. Along sandy washes where the soil holds more moisture, the light green paloverde, the dark green mesquite, and the pale green ironwood produce their clusters of beans.

There is considerable animal activity in the bleak expanses of the shrub community. Creosote bushes provide vital shade for the desert tortoises. Their underground burrows, which offer a retreat from the summer heat and a place to hibernate in the winter, are stabilized by the root systems of the bushes. The young tortoises are hatched from eggs and receive no care from the adults. These soft-shelled hatchlings, in great danger from predators, must learn to hide among the shrubs.

When water is available, the tortoise will drink its fill, and its hard shell and leathery skin will help preserve the supply. Tortoises feed on annual wildflowers and grasses in the spring and, when available, in the fall. Generally, the spring nourishment must carry the tortoise through the summer, the fall, and hibernation in winter, during which it must depend on the fat and water stored within its body. Among the other animals in the area is the round-tailed ground squirrel. It may emerge from its underground burrow early in the morning to gather the tender buds of creosote bushes. Like the tortoise, it must do most of its feeding after the rains in winter and early spring, when the supply of green food is most plentiful. Emaciated after a winter's sleep, the squirrel quickly fills its body in the short season of plenty. Although its tan coat blends well with the sandy surroundings, its sluggish movements after its feast makes it an easy prey for hawks and coyotes. The dwindling of the food supply triggers a period of estivation or hibernation. The crested lizard and other sand lizards are also residents of the shrub community, associating close by the rodents. When necessary these animals take refuge in the burrows dug by rodents around the base of the bushes.

The bajada, the area between the level plain of the desert and the foot of the mountains, is a favorite habitat of the cacti. The chollas predominate, especially the chain-fruit and teddy-bear cacti. Beneath these, smaller species such as barrel and hedgehog cactus and pincushion may be growing. Seeds of these cacti are scattered in the droppings of birds that roost among the sharp protective needles of the chollas. As dangerous as these needles are to humans, they seem to make ideal nesting places for a number of birds, including the roadrunners, the thrashers, and the cactus wrens. Each builds a different kind of nest set deep in the spiny joints where the young birds can have maximum protection from predators and be shielded as well from direct sunlight. The nest of the cactus wren, constructed of small twigs and grass stems, has been called a "purse" because of its elongated round shape. The cactus wren, the largest of the wrens, is recognized by its white eye-stripe, spotted breast, and white-barred tail. It forages for most of its food in the vicinity of the cacti. Leconte's thrasher, a pale-gray bird, feeds mostly on the ground in open cactus deserts, using its long downcurved bill as a tool to dig out ground-dwelling insects. Its nest, hidden in the top branches of cacti, mesquite, or paloverde, is an open structure of twigs lined with grass.

The largest nest of all, shaped like a shallow saucer and crudely constructed of large twigs and sticks, belongs to the roadrunner. This slender, heavily streaked, odd-looking cuckoo is constantly in a hurry, preferring to travel on the ground and to use his long tail and short wings to achieve fast stops and quick turns in escaping from enemies or in pursuing its food. It takes to the air only in the most demanding emergency. Its diet is extensive; it will eat cactus fruit and wild berries but spends most of its time patrolling the shrubby desert areas for lizards, birds, rats, worms, and insects. In an encounter with a rattlesnake, the roadrunner ultimately prevails because of its skillful footwork. The roadrunner dances around the snake, jabbing at its head and inciting the snake to expend its venom and energy in a continuous series of useless strikes. In the end, the roadrunner will stun the snake with its powerful beak, grab it below the head, and whip it against the ground until it is lifeless, and then drag it off to a shady spot to be swallowed whole and digested at leisure.

A common resident of the gravelly foothills is the antelope ground squirrel. The *antelope* part of its name comes from its white-backed bushy tail, which is often curled over the animal's body as it darts back and forth among the scattered brush. Because of the white stripes on its dark-gray body, this small ground squirrel is often mistaken for a chipmunk. Actually, it is related to the round-tailed ground squirrel of the open desert plains. As already noted, the round-tailed species is dependent primarily on the springtime supply of green foods and must become inactive when this supply dwindles. The antelope ground squirrel, whose scientific name is a combination of the Greek words for "sand," "seed," and "lover," feeds voraciously on the seeds of cacti, yuccas, and other plants as well as on leafy green foods and a variety of insects. It has no difficulty climbing over barbed cacti, thorny shrubs, or spiny trees, and when foraging will travel substantial distances from its burrow. Food is available for this lively animal all year round, so that its burrow is used only to escape from excessive heat or cold and for nighttime protection from predators, and rarely for estivation or hibernation. Other inhabitants of the rocky cactus areas include shrews, rats, lizards, and jackrabbits and the snakes, foxes, and coyotes that feed on them.

Jutting up from the vast expanses of the desert floor are barren mountain ranges, hot and dry, offering to plants and animals a harsh and desolate environment. This is the home base of the desert bighorn sheep, which seek out the solitude of the highest inaccessible ridges. They survive in this habitat, jumping from rock to rock and climbing rugged slopes, because their feet are adapted to grasp uneven surfaces and to give traction on smooth ledges. They were once abundant, but disease, habitat destruction, and trophy hunting have reduced their numbers to a precious few, most of which are now in protected areas. At best, life in these high mountain slopes is demanding. Food and moisture are never plentiful, and for a good part of the year both must come from the grasses, wildflowers, and leafy foliage of desert shrubs on which the sheep subsist. When the blistering summer arrives and the meager waterholes dry up, the sheep will use their horns and hoofs to rip open large cacti so that they can eat the juicy pulp. As distinguished from the bighorn sheep of the Rockies, which are darker in color, the desert bighorn take on the lighter grayish-brown shades of their rugged terrain. It is not easy to spot one at a distance, but when outlined against the sky, the bighorn becomes the magnificent animal of the mountains, with its massive coiled horns and its creamy white rump. During the mating season, the horns are weapons of attack as the rams fly at each other in head-on clashes that often stagger both contestants. The females' horns are smaller and curve only slightly backward. Predators are not a major problem. With their keen eyesight and sure-footedness, mountain sheep can outmaneuver their natural enemies. Despite the austerity of their environment, their numbers are increasing in the managed ranges of national and state parks.

In the western portion of the Mohave Desert, just inside California on its border with Nevada, is Death Valley, the lowest and hottest spot in the United States. The geological formation of this region is known as a *graben*—a depression resulting from the fracturing and uplifting of the bordering bedrock and the sinking of the area between flanking mountains. Death Valley is boxed in on all sides: the western wall of the valley is formed by the Panamint Range, with elevations between 6,000 and 11,000 feet; the eastern side is the Amargosa Range, rising to heights from 4,000 to 8,000 feet; the northern and southern ends of the valley are also blocked by mountain ranges. The valley covers approximately 3,000 square miles, measuring about 150 miles long and no more than 20 miles at its widest point. At one time the valley contained a vast lake; this evaporated as the climate became more arid, leaving behind salt flats hundreds of feet thick over an area of 200 square miles. Almost 500 square miles of the valley are below sea level, with the lowest point near Badwater reaching a depth of 282 feet.

The little rain that falls rushes down the mountainsides, only to evaporate quickly in the valley. From the surrounding mountains, fine particles of sand are washed into the basin and driven by the wind into dunes that are one hundred feet high near Stovepipe Wells. Unmerciful heat dominates the scene for six months of the year and relaxes its grip slightly for the next six months. The overall impression is that of a huge, monotonous expanse of desert bleakness. There are badlands, saltwater springs, and dry lakes. But with elevation, moisture increases until on the high peaks there are forests with juniper, mountain mahogany, and piñon pines. Between the two extremes is a wide

spectrum of desert life-forms. Only the salt flats are totally barren. The little rain that falls yields spring-
blooming wildflowers that can, in a favorable year,
briefly transform the desert into a vast garden. Cacti
and other xerophytes live where conditions permit.
Birds, lizards, snakes, night-active rodents, and bighorn
sheep confirm Death Valley's association with the Mohave Desert.

THE GREAT BASIN

The largest of all the American deserts is the Great Basin Desert, an area of approximately 200,000 square miles that includes all of Nevada, most of western Utah, and parts of California, Wyoming, Idaho, and Oregon. Such a vast region defies a simple description. In the main, it is a broad, flat, arid plateau almost completely locked in by tall mountain walls. The Rocky Mountains are on the east. On the west is the Sierra Nevada–Cascade barrier, which effectively cuts off the moisture-laden winds from the Pacific Ocean. By the time these winds cross the barrier, they blow warm and dry over the Great Basin. Much of the region is bleak, particularly in the north, where the terrain has the characteristics of a steppe. The southern portion takes on the qualities of a true desert as it gradually merges into the Mohave.

The floor of the basin is two to three thousand feet in elevation. Above it rise ranks of low mountain ranges, most no more than seventy miles in length, that run more or less parallel in a north-south direction. A few reach elevations of eleven thousand feet above sea level. All of these are fault-block mountains, thrust up millions of years ago by volcanic forces along fractures in the crust of the earth.

Rainfall in the Great Basin averages somewhat less than ten inches a year, and some snow falls at higher elevations. While the winters are cold, the summers are not always hot. The dominant vegetation is sagebrush, with shad-scale scrub taking over in the more saline habitats. There are few of the cacti and desert trees found in the Mohave and Sonoran Deserts, where the sun burns cruelly most of the year. To distinguish it from the deserts in the warmer latitudes or at lower elevations, the Great Basin is often referred to as the "high" or "cold desert." This may appear to be a contradiction in terms, but if the test for a desert is limited rainfall, then the Great Basin qualifies.

Hemmed in on all sides by steep mountain walls, except in the extreme north, which is drained by the Snake River, there is no drainage leading out of the Great Basin and very little leading into it. The north-south mountain ranges, spaced at intervals, divided the region into an overall pattern of highs and lows. Torrential rains in their downhill rush deposited large quantities of sediment at the base of the mountains. These deposits of sand, gravel, and clay built up into ever-widening fans and outwashes that in time grew large enough to coalesce into alluvial plains. When these met across the valleys, they formed barriers that divided the land between the mountains into a chain of basins devoid of drainage outlets. Now rainwater is trapped in these basins. Some filters into underground reservoirs, but most collects in the lowest part of the basins to form temporary lakes known as *playas*. In the heat and dryness of spring and summer, the water evaporates, leaving a residue of minerals carried down from the mountainsides. These are the mineral flats, conspicuous features of the desert landscape, which sparkle when seen from a distance. In some areas, the runoff flows into lakes left behind as remnants of the huge ancient bodies of water that once covered thousands of square miles of the basin region.

There is more to the Great Basin than a series of low mountains wriggling across wide plateaus and forming broad depressions for the collection of rainwater. In the southern part of Oregon, buttes and mesas rise above the desert floor. In the east, the Great Salt Lake and the extensive Salt Lake Desert dominate the landscape. Just on the western edge of the basin, inside the California border, is Death Valley, the hottest, driest, and lowest spot in the United States, formed as a result of the sinking of the earth between uplifted mountains. In common with other arid areas, there are spectacular sand dunes, the largest located in northern Nevada. There are snow-capped mountain ranges in the central region that are topped with thick greenery because their peaks receive an exceptionally heavy rainfall. There are bare and rocky places, and vast sandy spaces. Scattered throughout the basin are lakes kept fresh all year round by adequate inflow and constant drainage. The whole region is a complex of environments, but is treated as a unit because its geographic area is so clearly defined by the surrounding mountain systems. Even the Painted Desert in northeast Arizona, whose colors change rapidly with the varying weather patterns, and its forest of some of the largest petrified logs in the world, fall within the borders of the Great Basin.

A striking feature of the vegetational pattern of the Great Basin is the dominance of the various species of the sagebrush plant. They spread in profusion, mile after mile, through the broad basins, over gentle alluvial slopes, and to the summit of some of the mountains. Though they thrive best in well-drained, nonalkaline soils, some species manage well in moderately saline areas. They flourish everywhere, except in the driest of the flats and the highest of the peaks. Very few areas of the West are without these shrubs, but the Great Basin is legendary sagebrush country, where these low, fragrant, gray-green bushes flaunt their unchallenged dominion over the plant life of the entire region.

In the areas of the mineral basins, and in the drier sites where the soil is less penetrable, the sagebrush flats give way to the group of low-statured shrubs known as shad scale scrub. These are also called saltbushes, a name that alludes to their tolerance for saline environments. In these areas, shad scale crowds out other plants, producing pure stands of monotonously gray vegetation. Few plants can compete under the handicap of a soil burdened with salt. One of the most common of the shad scale plants is the four-winged saltbush. Its narrow, oblong leaves are covered with small silvery-gray scales that give this freely branched bush the characteristic unkempt appearance of this community. Newly grown tender leaves and shoots make a ready-salted, palatable vegetable salad. The flowers are inconspicuous, but the four-winged fruits are produced in prominent rose-colored bunches. Its seeds are ground

THE GREAT BASIN

PIÑON-JUNIPER

SHAD SCALE

BLACK-CHINNED HUMMINGBIRD
(Sometimes catches insects like a flycatcher.)

BLACK-THROATED
SPARROW

SHAD SCALE

SPOTT
SKUNF

DESERT TRUMPET

SALTBUSH

*Four-winged fr
are ground and m
with water and suga
make a drink called pi*

KANGAROO RAT

SOUTHERN
GRASSHOPPER MOUSE

HORNED "TOA
(Lizard)

LEOPARD LIZARD

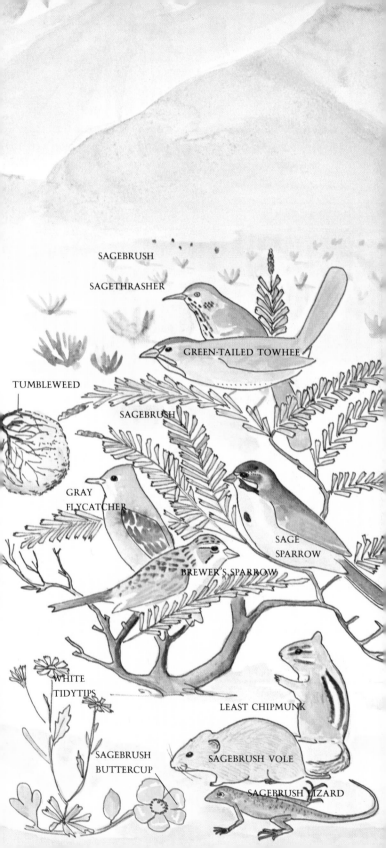

SAGEBRUSH

SAGETHRASHER

GREEN-TAILED TOWHEE

TUMBLEWEED

SAGEBRUSH

GRAY
FLYCATCHER

SAGE
SPARROW

BREWER'S SPARROW

WHITE
TIDYTIPS

LEAST CHIPMUNK

SAGEBRUSH
BUTTERCUP

SAGEBRUSH VOLE

SAGEBRUSH LIZARD

and mixed with water to make the drink called *pinole*. The plant is a valuable food for browsing animals. Another member of this community is the spiny saltbush or spiny shad scale. While similar in appearance to the four-winged saltbush, it has spiny twigs and flat fruit.

Sometimes growing together with the saltbushes, but often found in large colonies in wet alkaline soils, is the spiny shrub called greasewood. It has a smooth white bark, fleshy, wormlike, bright-green leaves, and fruit that looks like a miniature smoker's pipe. This also is a valuable plant for browsing sheep and cattle. The name *greasewood* is often applied to other plants, especially the widespread creosote bush of the Mohave Desert.

The most barren areas of the Great Basin are the low points in salt flats and alkaline sinks. The mineral concentrations in these locations are generally too high for any plant life, and even torrential rains are unable to leach the soil of these desiccating deposits. As time goes on, the periodic evaporation of the waters that temporarily stand in these depressions merely extends and increases the unproductiveness of the environment.

On the whole, the Great Basin is a colorless region almost devoid of trees. But some of the mountain ranges are high enough to support sporadic forests of coniferous trees. On the highest peaks, covered with snow during the winter, there are even some specimens of bristlecone pine, the elders of the summits, standing twisted and weathered after thousands of years of growth. On the lower mountain slopes, at elevations generally from five to seven thousand feet, growths of piñon pine and juniper occur, the same type of pygmy-shrub community encountered in southern Utah, parts of Colorado, and northern Arizona and New Mexico. Along some of the watercourses, cottonwoods and willows manage to grow. All of these add spots of color to the dry and barren landscape. Unseen are the deposits of gold, silver, copper, zinc, lead, and uranium. These are mined throughout the Great Basin—a land rich in the lore of boomtowns, land claims, and tireless prospectors.

Much of the animal life in the Great Basin is similar

to that found in the adjacent areas of the Rockies and in the other southwestern deserts. Mule deer, pronghorn, and elk feed during the winter months in the valleys and on the piñon-juniper slopes. There are many species of rodents in the sagebrush and saltbush communities, including wood rats, pocket mice, voles, chipmunks, and kangaroo rats, some of them differing from the species in the warmer deserts. Black-tailed jackrabbits and mountain cottontails are also abundant in this sparsely vegetated desert, moving alertly from bush to bush to escape their predators. Among the carnivores are the spotted skunk, badger, mountain lion, and the ubiquitous coyote. Insects and spiders are preyed upon by the sagebrush lizard and the leopard lizard. In the northern region of the Basin, where there is more vegetation and more rainfall, there are short-tailed weasels, raccoons, and red foxes. Desert bighorn sheep browse among the vegetation on the highest mountain crags and at times descend to feed on the sagebrush. Most of the bird species of the Southwest, except those in the lowest and driest areas, are represented. The most spectacular are the vast colonies of shore birds that nest on the islands of the salt lakes in Utah and Nevada. The enormous freshwater marshes in southwestern Oregon support huge populations of ducks, coots, herons, and ibises.

The variety of animal life reflects the spaciousness of the region and the diversity of its communities. But in all, the Great Basin is an unyielding place, with barren mountain ranges, dry lake basins, salt-encrusted playas, vast sandy spaces, alkaline sinks, and brush-covered plains. The three large cities in the region—Salt Lake, Reno, and Las Vegas—barely fall within its boundaries. Fresh water and tillable soil are at a premium, and areas habitable by man are mimimal.

SALT LAKES AND FLATS

The "modern" history of the lakes and flats in the Great Basin region began about 70,000 to 100,000 years ago, during the last stages of the Ice Age. For millions of years before that, the land between the Rockies and California was being formed and shaped through invasions by the sea, the depositing of limestone beds, the folding and faulting of the earth's crust, the elevation of mountains, and the exposure of salt-bearing sedimentary rocks. The Ice Age brought with it damp and chilly weather. During the warmer periods, rainwater and meltwater were released in massive quantities. The vast region of the Great Basin was a convenient receptacle, especially where mountain ranges encircled depressed areas. Under these pluvial, or rainy, conditions, the waters drained into these areas for thousands of years, forming numerous freshwater lakes that stretched for miles between the surrounding mountains. The two largest of these prehistoric lakes were Lake Bonneville, which covered much of northern Utah and parts of Idaho and Nevada, and Lake Lahontan in western Nevada and part of California. But the weather during these times was not uniform and the glaciers were not static. Periods of drought followed periods of moisture with consequent changes in the size and levels of the lakes.

By studying the contours of the mountains surrounding these lakes, geologists have determined their original dimensions as well as the various changes in their water levels or shorelines. Many of these physical features, especially terraces or benches, worn into the flanks of the mountains by wind and wave action during long periods when the lake levels were unchanged, are plainly visible even to the casual observer. Some of these, as in the Wasatch Range in northeastern Utah, resemble abandoned roads carved into the mountain rock, and are key features in determining the progress of Lake Bonneville's shrinkage into three smaller lakes as the glacial waters receded.

The largest of these remnants is Great Salt Lake, "great" in the sense that it is the largest lake in the United States west of the Mississippi, but after three major declines in its level, the lake is now only a small fraction of the size of its predecessor. At its highest and widest level after the glacial melt, Lake Bonneville was 100 miles wide and almost twice as long, and reached a depth of more than 1,000 feet above the present Great Salt Lake. This is known as the Bonneville Terrace. In time, the lake started to run over the rim of the Great Basin at Red Rock Pass, draining off into the Pacific Ocean by way of the Snake and Columbia rivers. The waters rushing through the pass cut deeper and deeper until they finally reached bedrock. The lake had dropped by about 375 feet, or 625 feet above its present bed, to what is known as the Provo Terrace. The lake stabilized and remained at this level for an extensive period of time, until periodic changes in climate began to reduce the inflow of water to a point where it did not replace the amount being lost through evaporation. The lake shrank to its third level, the Stansbury Terrace.

The level of the lake rose and fell for years, but as the climate grew warmer and drier and the water supply diminished, the trend was steadily downward. At present at its deepest point, along the center section, the lake is only thirty-four feet deep, with an average depth that is much lower. In the last hundred years or so, the level of the lake has fluctuated within a range of about twenty feet. Today it is almost at the level at which it stood in 1850 when record-keeping began, indicating that during this period the rate of evaporation on the whole has been more or less balanced by the rate of inflow and precipitation. Today, Great Salt Lake covers an area of approximately fifteen hundred square miles, a miniature of its original size, as are the two other lakes left behind by the contraction of Lake Bonneville. These two are Sevier Lake and Utah Lake. The latter, a freshwater lake fed by the snow-covered mountains of

the Wasatch Range, remains fresh because it has an outlet by way of the Jordan River into Great Salt Lake.

It is the lack of an outlet to the sea that accounts for
the salinity of Great Salt Lake. When it finally reached
the point at which no more water flowed out over the
Red Rock Pass, the lake lost the beneficial effects of a
constant freshwater bath. Millions of acre-feet of fresh
water are annually washed down from the surrounding
mountainsides, and these carry along with them in solution minute quantities of minerals drawn from the
weathered rocks over which the waters flow. These
rocks are the product of the material laid down during
the years of invasion by the sea.

As long as there is an outlet from a lake, minerals in
solution are washed away and the lake can remain
fresh. But when there is nowhere the water can go, huge
quantities evaporate from its surface, and the lake must
hoard the dissolved minerals year after year. These
minerals consist mainly of sodium, magnesium, po-
tassium, lithium, and calcium, and these combine with
other substances to make chlorides, sulfates, and carbonates. Lake waters are considered salty when the
dominant minerals are chlorides or sulfates. They are
classed as alkaline when carbonates prevail. The accumulation in Great Salt Lake, which started thousands
of years ago and is still going on, has resulted in an ever-
increasing concentration of minerals. About eighty-five
percent of the mineral content of the lake is sodium
chloride, or common salt. The present lake is approximately twenty-five percent salt, or roughly one part salt
for every three parts of water. By comparison, ocean
water is one part salt to thirty parts water. There is
always speculation whether the Great Salt Lake will in
time evaporate completely. Since evaporation decreases
as the lake becomes saltier, it will take a prolonged drying trend to turn the lake into a salt flat.

With so great a concentration of salt, fish cannot survive. Dead fish are seen occasionally, but these have
either come in from inflowing streams already dead or
died shortly after their entry into the lake. The salinity
of the water severely restricts all forms of life, much of

which is microscopic. The most fascinating visible life-form is the tiny brine shrimp, an animal that is rarely more than one-half inch in size. It is a black-eyed, pinkish creature with five pairs of bristly legs, existing in such enormous quantities that the water sometimes appears to have a pinkish tint. A cup dipped into the water at the edge of the lake will bring up several of these transparent creatures, which are so numerous they are harvested commercially as a tropical-fish food. The eggs of these shrimp can be stored until needed, and hatched within a few days by dropping them in salt water. There are also several forms of algae. The green type is a basic food for the brine shrimp, and a red type, heavily concentrated at the north end, where the salinity is highest because it receives little fresh water, gives a red tint to the lake.

Brine flies match the brine shrimp in numbers. Millions of them form small black hoards that move like sheets over the beaches and lake debris. The larvae hatch into water-borne wrigglers, become pupae and finally emerge as flies. While these flies are minor pests to bathers, the pupae shells that collect on the shore foul both the beaches and the air. Horseflies, deerflies, gnats, and mosquitoes are the usual nuisances to man and beast.

The salinity of Great Salt Lake has made it an attraction as a bathing resort. Swimmers are unable to sink, and the sensation of popping up in its buoyant water and floating effortlessly attracts thousands of tourists. There are some minor hazards attached to this pastime that may come from swallowing a mouthful of water or getting some of it in the eyes, and from the irritating effects of salt crystals on the skin when exposed to the air. Most tourists who want to bathe now visit the state park and beach on Antelope Island, a peninsula jutting into the lake from the south. There is boating on the lake by sailors who have learned to cope with the corroding effects of so briny a solution. The millions of tons of salt and the other minerals in the lake have stimulated a mineral extraction business. Lake water is pumped into artificial ponds and allowed to evaporate

WHITE PELICAN *nests in salt lake but feeds in fresh water.*

LONG-BILLED CURLEW

BLACK-NECKED STILT

DESERT SALTGRASS

Smaller than MOSQUITOES, MIDGES *are the most important food for birds. Commonly mistaken for* MOSQUITOES, *they emerge throughout the summer bu are most common during May, June, an September. Large black clouds are form by mating swarms in spring and fall.*

HORSEFLY

MIDGE

HORSEFLIES *and* DEERFLIES *create a serious nuisance for wildlife, livestock, and man.*

BRINE SHRIMP

CALIFORNIA GULL

SALT LAKES AND FLATS

WESTERN GREBE

AVOCET

WILSON'S PHALAROPE
(more colorful female)

SLENDER
GLASSWORT

BRINE FLIES *Larvae hatch into waterborne wrigglers, then change into baglike pupae fastening to shore and lake debris.*

BRINE SHRIMP's *five legs look like many more as they rapidly paddle in undulating waves.*

BRINE SHRIMP

MOSQUITOES
An abundant pest

in the sunlight. The process is repeated until a thick sediment is collected, which is then removed, processed, and marketed.

The huge prehistoric lake in the western part of the Great Basin, Lake Lahontan, has also left a string of smaller descendants. As in the case of Lake Bonneville, distinct terraces exist that indicate the former levels of this lake, and some of these are levels believed to reflect the ends of various glacial periods. Lake Lahontan was also a freshwater lake that began evaporating when the final glaciers retreated and the torrential rains ceased. It has shrunk considerably from its one-time expanse of some 8,000 square miles and has left behind, among others, Honey Lake in California, and Pyramid and Walker lakes and Humboldt and Carson sinks in Nevada. Because of its drainage outlets, Honey Lake has been kept fresh, and Pyramid and Walker lakes are only slightly saline. Pyramid Lake, Lahontan's most notable descendant, now covers an area of 168 square miles, one fiftieth of its original size. Some of these remnant lakes, such as Mono Lake in California and Soda Lake in Nevada, qualify as alkaline lakes because of the high concentration of carbonates in their waters. Numerous other smaller lakes are scattered over western Nevada. Many of these are little more than mineral flats that are covered with a few inches of water for part of the year. Some of these lakes disappear completely during long periods of limited rainfall.

There is spectacular bird life on and around these lakes. Rookeries of herons, egrets, stilts, and other shore birds exist on the various islands of Great Salt Lake. Food is generally found miles away in the mud flats, marshes, and open waters at the eastern fringes of the lake. Refuges have been established there, and these also accommodate ducks, geese, swans, and other migratory birds. The rookery on Gunnison Island supports a large colony of white pelicans that fly to these refuge areas, feed voraciously, and return to let the young pelicans reach into their bills for a regurgitated meal. Unlike brown pelicans, which dive into the water for food, white pelicans swim about and scoop up fish in their bills, often moving along in arclike flocks. The pelican rookeries always face the danger that a drop in the level of the lake will create sandbar passageways for such land-based predators as raccoons, coyotes, and rattle-snakes. Seagulls are everywhere, and while they do some damage to the eggs in these rookeries, they are protected by state law and have been honored by a monument in the heart of Salt Lake City. Erected by the Mormons, it pays tribute to the thousands of gulls that saved the early settlers from possible starvation by devouring a horde of crickets that threatened their crops. Unlike other birds in the region, the gulls do feed on the brine shrimp and flies in the lake. A bird refuge, primarily for pelicans, has been established on Anaho Island on Pyramid Lake. Mono Lake has the same brine shrimp and black flies as Great Salt Lake, and similar congregations of shore birds. In addition, other water birds also nest on various islands in the lakes in the Lahontan region.

Scattered between the ranges of the Great Basin are large, flat areas that are covered by water only during periods of heavy rain. These are the remnants of bodies of water that have completely dried up during the ten thousand years since the end of the last Ice Age. Many of these areas are rimmed by shoreline terraces left behind by the ancient salt and alkaline lakes. Without drainage outlets and with minimal percolation through their surfaces, the mineral accumulations in these locations have increased year after year as standing water periodically evaporates under the hot desert sun. The few basins that do reach ground water hold permanent ponds, bordered usually by crystalline deposits.

The most extensive of these dried-up areas is the three thousand square miles of desert east of the Great Salt Lake. Some two hundred square miles of this desert make up the famous Bonneville salt flats, covered with a thick layer of common salt. There is water in these flats only during winter and spring, and when the salt is blown about by the wind, it smooths the surfaces. When the water has evaporated, it leaves behind a salt pan almost as hard as concrete and so flat that it is said to be the only place in the United States where the curvature of the earth can be seen. This hard, flat surface, level over such a great distance, has become famous as an international automobile speedway. Races are scheduled in late summer, when the salt is exceptionally dry and firm.

In the span of geology, the salt flats are of recent origin. They make harsh and sterile environments. Their future depends on the weather. Like deserts, they will shrink during wet periods and advance into contiguous areas during long periods of drought.

SAND DUNES

The supply of sand for making dunes may come from the beaches of an ocean, the shallow margins of lakes, or from the sediments along the flood plains of rivers. In the West, where there are some of the most spectacular dunes in the world, the sand came down from the mountainside, grain by grain, as streams and rivers tumbled down and deposited vast quantities of debris in the valleys below. The dunes in southern Colorado, now preserved in the Great Sands National Monument, contain mainly volcanic rock fragments and bits of quartz from the components of the mountains that surround the San Luis Valley. Erosional forces were at work on these mountains for countless years, but when the arid climate developed after the end of the last Ice Age, loose sand was left behind in the basins of dried-up lakes and in the old riverbed of the Rio Grande. Sand is still being torn from these mountainsides as their slow disintegration continues under the impact of snow, frost, and summertime thunder showers.

In White Sands National Monument in New Mexico, material for the sand dunes came and still comes from the sedimentary layers of rock in the mountains that imprison the Tularosa Basin. Among these layers of rock, which disintegrate under desertlike conditions, are formations containing large quantities of gypsum, a mineral easily soluble in water. For thousands of years, gypsum and other minerals were washed in solution down onto the floor of the basin, part of which was then covered by a large lake. With the change in climate some ten thousand years ago and the coming of hot, arid conditions, much of the lake evaporated, leaving behind in its dry basin large deposits of gypsum. Under continuous weathering and abrasion, the deposits of gypsum crystals break up into fine grains, the material from which the dazzling white sand dunes are made. The present remnant of the ancient lake is a playa, filled with rainwater only during the summer showers and for a short period thereafter. The supply of gypsum keeps increasing year after year as the playa dries up under the heat of the desert weather.

The dunes on the western edge of Death Valley in southern California consist of fine fragments of sandstone and granite. These are the sediments of fluctuating streams that flow into the basin floor from the surrounding mountains. In mountainous areas, the dune-making process is always the same; the mountains themselves surrender the material for any sand dunes in the valleys below them.

Sand alone will not make a dune. What is needed in addition is a brisk steady breeze and an obstacle in its path. On desertlike floors where the vegetation is too sparse to hold down the sand, winds sweep across the basin or valley, bouncing and rolling the loose sand until it is stopped in its path by a rock, a plant, a clump of grass, or any other obstacle. The wind-powered movement of grains of sand, a few inches at a time and rarely rising far above the ground, is called *saltation*. As this sand is trapped, it backs up and forms a mound, which, once started, becomes its own obstruction, forcing the accumulation of more and more sand. The dune grows as grains of sand bounce up the windward slope, building a heavier and heavier crest. The sand driven up the crest on the windward side slips down on the lee side of the dune when the so-called angle of repose is reached. This is the steepest angle at which a pile of sand will hold without collapsing. This classic profile of dune has a gradual, gentle windward slope created by the upward movement of saltating sand, and a steeper slipface on the leeward side caused by the avalanching of the crest.

This process of saltation and avalanching accounts for the migration of a dune in the direction of the prevailing wind as material is continuously carried from the windward to the leeward side. A reversal of the wind pattern will carry the dune, layer by layer, in the opposite direction. Where the direction of the wind changes regularly or where it comes from various directions, dunes remain in fairly confined areas. Some dunes, especially those at seashores or lakesides, are often anchored in long-term positions by deep-rooted vegetation. But many dunes travel in one direction, often burying shrubs and trees that are later left behind and exposed as pathetic ghosts when the dunes continue their advance.

The only certainty about the contour of a dune is that it will change. The wind blows, grains of sand keep bouncing, and the dune begins to take on a new shape. The resulting structural patterns depend upon such factors as the supply of sand, the direction, duration, and velocity of the wind, the density of the surface vegetation, and the general topography of the region. There are several recognizable dune patterns, each created by a different combination of these factors. Transverse dunes are constructed when large quantities of sand are available. Intervening vegetation is buried, so that the crest rolls along in a fairly straight line at right angles to the wind's direction. Crescent-shaped dunes are usually formed where the winds are moderate and the amount of sand is limited. The milder air currents cannot sweep over the crest of the dune. Instead they verge around it, depositing the sand in a curving pattern with the tips on each side jutting forward. This type of dune is called a *barchan*, an Arabic word meaning "ram's horn." Barchans have gradual slopes on the windward sides and steep-sided walls on the leeward sides, with the sharp ridge undulating in a series of crescents. The reverse of the barchans are the parabolic dunes. These occur when firmly rooted vegetation holds down both sides of a dune, so that only the center can curve in the direction of the wind. The tails on both sides of the nose taper off in the opposite direction.

Changes in wind direction can result in special structural patterns. The seasonal shifts in the winds at Great Sand Dunes National Monument account for the enormous height of these dunes. The movement of these transverse dunes in one direction during one part of the year is canceled by winds in the opposite direction during another part of the year. Windward slopes become

SAND DUNES

A ram's horn dune ("Barchan")

Dead tree formerly covered with sand and now being exposed again as dune moves.

KIT FOX, *a visitor to the dune*

PRICKLY POPPY

A few grasses grow in the interdunal hollows, the "swale."

SIDEWINDER tracks are at right angle to direction of movement.

SIDEWINDER

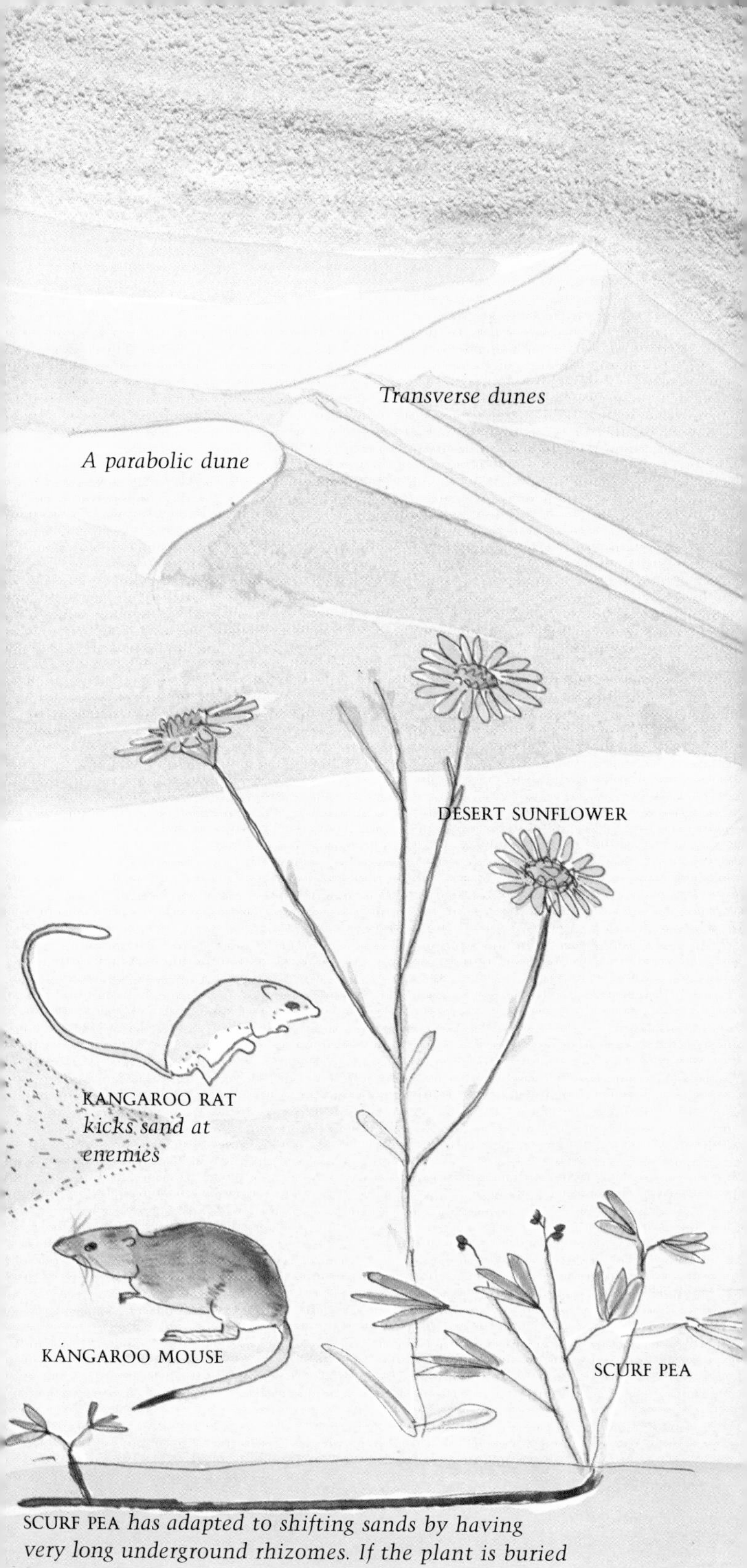

Transverse dunes

A parabolic dune

DESERT SUNFLOWER

KANGAROO RAT
*kicks sand at
enemies*

KANGAROO MOUSE

SCURF PEA

SCURF PEA *has adapted to shifting sands by having
very long underground rhizomes. If the plant is buried
by sand in one place, it can sprout at another.*

leeward slopes as the direction of the winds change, with the result that the dunes just keep folding back on themselves, with new sand having nowhere to go but outward and upward. These are sometimes called *reversing dunes*. A change in the direction of the wind may create a small reversed dune on top of the principal dune, so that small slipfaces wind along the crest in what is often described as "Chinese walls." Sometimes breezes sweeping across the windward slopes form thousands of ripples at right angles to the wind. These tiny crinkles, varying in pattern from day to day, add shadow and texture to the mounds of sand.

A sand dune is a formidable environment for any living thing. The sand is sterile. Food is hard to come by. Rainwater sinks quickly below the surface. The dunes heat up to lethal temperatures during the day and cool down with amazing suddenness at night. The winds are dry, blasting and dehydrating everything in their path and often uprooting plants that have managed to gain a foothold in the shifting sand. In some dunes the percentage of salt is too high for most vegetation. As in other hostile areas, the sand dunes have had their share of evolutionary failures. But a sparse population of plants and animals have adapted to existence in the region and continue to thrive despite the severe challenges of the environment. Most of the life begins with pioneer plants that become established in the flats between the dunes. These provide food and shelter for such full-time residents as seed-eating rodents and a variety of insects. The local food web is set in motion.

The problems of heat and dryness for plants among the dunes are the same as those faced in most desert environments. Nutrients are limited. Water below the surface of the dune must be reached by means of long taproots. Moisture must be conserved by such means as reduction in leaf size, waxy coatings, or fine hairs that reflect sunlight. On the dunes themselves the major enemy is the advancing sands. Several plants, in different areas, have solved this problem with the same unique "stem-stretching" adaptation. One of these, the Spanish bayonet, or soaptree yucca or little yucca, speeds up its rate of growth, elongating its stem to keep its flowering head above ground as threatening sands pile up around its base. In addition, the buried portion of the stem keeps shooting out more and more roots at successively higher levels. These are called *adventitious* roots, and their networks bind down the loose sand, improve the plant's capacity to soak up moisture, and, as they decay, add organic matter that nourishes other plant life. Not all of these yuccas survive, but when the sand moves on and uncovers those that have, their long, slender stems and network of roots are left exposed and helpless, ready to collapse in the blistering sunlight. Another stem-stretching plant is the skunkbush sumac, named for the pungent odor emitted by its leaves. This woody shrub is broad-based to begin with, so that with the addition of its adventitious roots it develops a foothold that can support it in place for many years. In the creek bottoms, narrow-leaved cottonwood trees often appear to be stunted in their growth. They too defend them-selves against suffocation by extending their trunks above the advancing sands and developing adventitious roots. Trees cannot just stand and wait. A pine buried up to its crown may be just a bleached skeleton when finally exposed by a dune that has moved on.

Some plants have evolved horizontal root systems for survival in the moving sand. The roots of the sweet-smelling rosemary mint grow along the ground. When they are buried, their tips bend and rise upward to get above the sand. The scurf pea defies sandstorms by sim-ply moving out of the way. Its long, horizontal stem, or rhizome, is not on the surface but remains buried un-der the sand. Rootlets can grow all along the rhizome. If a plant at one point dies of suffocation, the scurf pea sprouts new roots and new shoots at another avail-able spot.

Grasses may be seen in the more stable hollows be-tween the dunes. These are by far the most powerful protectors of the soil against the erosive forces of wind and water. The roots of Indian rice grass have a special stabilizing effect. They secrete a sticky substance that binds the grains of sand and holds them in place. The drooping blades of blowout grass shift with the winds and trace circular patterns in the sand around them. In this otherwide drab environment, the scene occasionally comes alive in midsummer with the sparkle of the blossoms of prickly poppy and desert sunflower on the uppermost reaches of the dunes.

Life among the dunes is as challenging for animals as it is for plants. The resident species comprise a handful of small creatures, and these few rarely venture far into the dunes. Survival depends on structural adaptations and behavioral patterns designed to cope with the niggardly environment. To escape the heat and dryness, much of the animal life is subterranean. Those that do not dig their own tunnels either use burrows abandoned by others or find relief in the shade of available vegetation. In the cold winters, snakes hibernate. In the dry summers, toads estivate. Some animals have evolved light-colored coats to match the surrounding dunes. Rabbits survive because of their enormous rate of reproduction. The kangaroo rat and the pocket mouse remain active without any water source, producing whatever liquids they need from their limited diet of seeds. Few of the animals can be seen during the day. For most of them, life is largely nocturnal.

The extent of this nighttime activity can be read by skilled observers early the following morning in the tracks and markings left on the surface of the sand. The sand may show the delicate string of dots made by the feet of beetles. On the way up a sand dune, lizards push aside little mounds of sand, separated by a thin path impressed by their tails. The sidewinder, a "horned" rattlesnake, is well adapted for traveling in loose sand. It moves by angular body motions, leaving behind a series of parallel diagonal depressions. The markings left by other snakes are long wriggling tracks. The delicate imprints of rodents may be traced across a dune. As the kangaroo rat hops across the sand its hind feet land simultaneously, leaving trails of double prints. Markings are also made by its tail as it keeps striking the ground in rhythm with the hops. Where two such trails meet, the sand may be scattered haphazardly, possible evidence of a territorial quarrel between two of these rodents. When threatened, the kangaroo rat will often kick sand in the face of a predator.

A visit by a predator is often marked by a sizable disturbance in the sand. Coyotes, horned owls, and kit foxes prey on residents of the dunes. The pad marks of a hunting coyote may show the impressions made by its four claws. The struggle between an owl and its victim will leave a nondescript jumble of sand. Deep hollows may be those of a fleeing jackrabbit. The two front pads of a kit fox may be identifiable, but the sand is an unclear jumble of footprints where the fox has outmaneuvered a kangaroo rat. Crosses two inches long may be the tracks of a roadrunner on the prowl. It may be difficult to isolate and identify the precise markings left by a particular animal, but the number of crisscrossing tracks and disturbed areas attest to considerable activity. By sunrise most of the animal life will be underground, and by midmorning the wind will have erased all evidence of this nighttime activity.

GEYSER BASIN

The earth's heat, geothermal energy, is what makes hot springs and geysers and such related phenomena as mud pots and fumaroles. Volcanoes occur when, as a result of enormous intense pressures, pockets of hot liquid rock, known as magma, erupt to the surface of the earth. In this process, water newly born within the earth's interior, called juvenile water, is thrust into the atmosphere. Hot springs and geysers are not considered volcanic because less than one percent of their water is juvenile water. Almost all of it is meteoric, water that has already been in the atmosphere and is returned to it in the form of steam or vapor only after having first seeped down into underground fissures. But the very existence of these hot springs and geysers generally indicates that the area in which they occur has, in terms of geological time, recently witnessed volcanic activity.

The most dramatic example of such an area is Yellowstone National Park. While it has been thousands of years since it witnessed its last volcano, the ten thousand or more thermal phenomena in this region exist because there is a body of magma below the surface that is still solidifying and discharging enormous quantities of heat. The cooling process gives off hot gases that rise toward the surface. If the magma is close enough to the surface that its escaping gases can reach the zone of ground water and react with it, the mixture will produce hot springs and geysers. The cooling process of the magma is so slow that in terms of human time the source of heat seems to be almost inexhaustible.

In all hot springs, the ground water rises from below the surface at an abnormally high temperature. Such terms as *boiling springs*, *hot springs*, or *warm springs* merely describe the levels of water temperature. Hot springs are found all over the world and have been used by man for bathing and washing since earliest times. There are hot springs in England, Germany, France, Italy, Switzerland, and various parts of the United States. Many of these places have names that include the word for bath, such as *Baden* in Switzerland, *Aix-les-Bains* in France, *Marienbad* in Germany, and *Carls-bad* in Czechoslovakia. The town of Bath in England was used as a health spa by the Roman conquerors. But not all hot springs are related to magmatic heat. The source of heat may be the friction caused by the movement of the earth's crust. Rocks heated in this way may retain their warmth for long periods of time. Another source may be chemical reactions, such as the oxidation of large bodies of mineral sulfides going on deep below the surface of the earth. The heat may also come from the pressure in the earth's interior. It is estimated that temperature in the earth rises an average of 1.8°F. (about 1°C.) for each 100 feet of depth. Even if this gradient is not constant, the temperature toward the earth's center may conceivably exceed the heat on the surface of the sun. The interior heat of the earth is one of the major problems faced by such relatively shallow operations as diamond mining in Africa or deep well digging in various parts of the world.

Whatever the cause of their high temperature, the heated waters of hot springs are chemically active and usually contain large quantities of dissolved minerals. As this water reaches the surface it cools and precipitates these minerals, many of which add brilliant blue and green colors to the pools. In the famous Mammoth Hot Springs area in Yellowstone National Park, the springs rise out of soft limestone beds buried hundreds of feet inside Terrace Mountain. The dissolved limestone is deposited as travertine encrustations that build up and spread out laterally in the form of flat-topped terraces. The bubbling water trickles down from terrace to terrace. Travertine makes a light porous rock and builds up quickly, at the rate of six inches to a foot a year.

Some of the colors in the open pools of hot springs are contributed by various algae that thrive in these

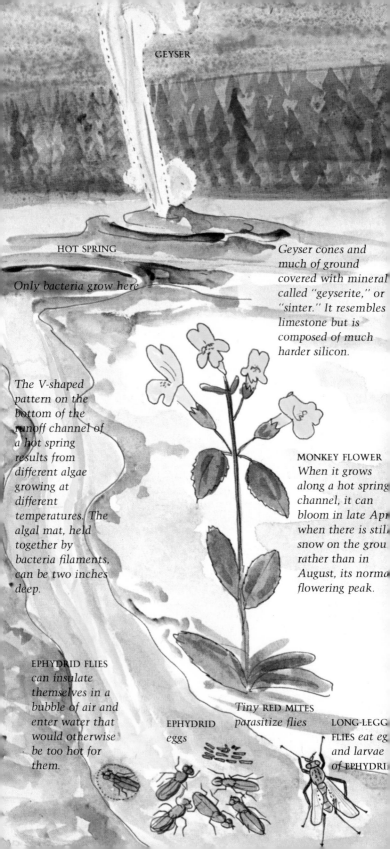

GEYSER

HOT SPRING

Only bacteria grow here

Geyser cones and much of ground covered with mineral called "geyserite," or "sinter." It resembles limestone but is composed of much harder silicon.

The V-shaped pattern on the bottom of the runoff channel of a hot spring results from different algae growing at different temperatures. The algal mat, held together by bacteria filaments, can be two inches deep.

MONKEY FLOWER When it grows along a hot spring channel, it can bloom in late Apr when there is stil snow on the grou rather than in August, its norma flowering peak.

EPHYDRID FLIES can insulate themselves in a bubble of air and enter water that would otherwise be too hot for them.

EPHYDRID eggs

Tiny RED MITES parasitize flies

LONG-LEGG FLIES eat eg and larvae of EPHYDRI

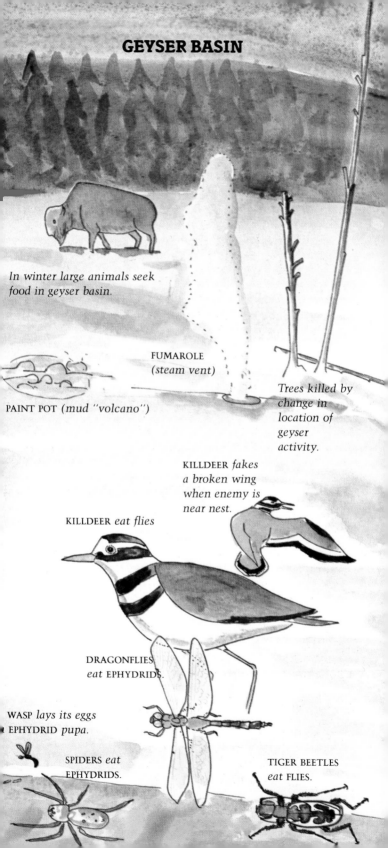

GEYSER BASIN

In winter large animals seek food in geyser basin.

FUMAROLE (steam vent)

PAINT POT (mud "volcano")

Trees killed by change in location of geyser activity.

KILLDEER fakes a broken wing when enemy is near nest.

KILLDEER eat flies

DRAGONFLIES eat EPHYDRIDS.

WASP lays its eggs EPHYDRID pupa.

SPIDERS eat EPHYDRIDS.

TIGER BEETLES eat FLIES.

scalding, mineral-rich environments. As the water flows off and begins to cool, bright-colored species of this microscopic plant cover the ridges and shallows of the runoff channels. Different-colored algae, some red, yellow, or brown, grow at different temperatures. The mat of algae may at times be two inches thick, held together by filaments of bacteria. A food web involving spiders, dragonflies, wasps, beetles, and even killdeer depends upon the curious ephydrid fly, which can enter the hot waters of these pools by insulating itself in a bubble of air. The killdeer, a bird of the open country, may be seen along with other birds in the areas warmed by the drifting vapors where it often nests. Near the nest, it feigns injury and hobbles along as if its wings were broken in order to lure a predator away from its eggs or young. Some bears have found steam-heated dens in these areas. In the wintertime the snow-free forage in the warm valleys attracts buffalo and other large mammals.

There are various gradations of hot-spring activity. Some just flow quietly, others emit escaping bubbles and boil over occasionally, and a few explode intermittently with a jet of water and steam. The explosive type is the geyser, and there are fewer of these because their periodic eruptions can only be produced by a unique underground plumbing system. The word *geyser* is Icelandic in origin and means "gusher." There are only three great geyser locations where tons of water and steam can be seen shooting out of the trembling earth. These are in Iceland, on the North Island of New Zealand, and in Yellowstone National Park. The Grand Geyser of Iceland subsided after having spouted for hundreds of years, leaving behind a huge cone of mineral deposits surrounded by a number of smaller active geysers. Iceland has about thirty true geysers, or about one percent of the total number of its hot springs. By comparison, the two hundred geysers in Yellowstone represent a much larger percentage of all its thermal phenomena, and these are by far the largest geysers, as well as the biggest geyser concentration, anywhere in the world.

The most famous of these is, of course, Old Faithful, not because it sends up the highest spout, but because its activity is both dramatic and predictable. Its uproarious discharge of ten thousand to twelve thousand gallons of water in a column from one hundred ten to one hundred fifty feet high lasts from two to five minutes and has occurred on the average of about every sixty-five minutes for the hundred years or so during which it has been under observation. The interval between explosions is governed primarily by the time it takes for enough water to sink back into the underground passageways and for it to be heated up to explosive intensity. Any abrupt change in the system of passageways, such as may be caused by a nearby earthquake, could disturb the "clock" that controls the regularity of the explosions.

There are geysers that spout higher than Old Faithful, but not with its regularity. There are some that erupt in the form of a fountain with a display suggesting fireworks, and others that emerge at an angle and form arcs across the sky. It is the combination of density and constancy that has enshrined Old Faithful as the goddess of the thermal world. But for a goddess it has a restricted life-span. Geysers are not forever. The cooling of the heat-bearing rocks and the erosive effect of flowing water on the underground fissures, however slow, will in time reduce a geyser to a boiling spring. Signs of deterioration have been observed even in Old Faithful. It is now about three hundred years old, and as it ages its eruptions may become less regular, less forceful, and spaced at longer intervals. In Iceland, the tendency of geysers to fade into inactivity is sometimes overcome by pouring soap flakes into the mouth of the lethargic pool. The scientific reason for this is not clear; but probably because of the disturbance in the delicate surface tension of the water, this practice "tickles" the pool into geyser activity. The same effect is sometimes achieved by using a stick to stir up the pool. As a conservation and anti-pollution measure, neither of these practices is permitted in Yellowstone National Park.

A geyser depends on the special features of its subterranean network of tubes. They must be irregularly shaped, so that water in these tubes cannot freely circulate and the normal equalization of heat by convection cannot take place. In addition the passageways must be so constructed as to permit an explosive discharge of the superheated water at the base of the column to exit through the single opening at the top. Finally, the whole structure must be strong enough to withstand the enormous pressures that build up in the system. The operation of geysers was first explained in 1880 by Robert Bunsen, the German chemist for whom the Bunsen burner was named. Though not applicable to all geysers, his generally accepted explanation depends on the nature of boiling water under pressure. The boiling point in a column of water increases with the depth of the column, so that under the weight of a column 500 feet high, water requires a temperature of 394°F. (201°C.) to boil instead of the 212°F. (100°C.) required at sea level. At some point, the superheated water at the base of the column begins to boil before the water above. This releases bubbles of steam that rise up the column, expanding as they float upward, and these force some of the water out of the column. This reduces the pressure in the column, setting off a chain reaction of lower boiling points. The entire column becomes unstable and flashes into steam with a tremendous explosive force that throws the whole mass out into the air. The process is repeated when water flows into the tubes and is again heated to the boiling point at its base.

Geysers, like hot springs, contain large quantities of dissolved minerals. Where the geyser waters rise through igneous rock, the minerals are harder than the limestone of hot springs. They are precipitated out slowly in the form of silica, known as geyserite or sinter, around the mouth of the geyser as a cone or in the form of wide, flat terraces. These deposits are also often brightly colored by the minerals themselves and by the various species of algae that flourish in these areas. The mound built up about the base of Old Faithful is more than twelve feet high.

Mud and fumaroles are two other thermal formations. The mud pot, also called the *paint pot*, occurs when steam rises to the surface through a pool of mud. The bubbles sometimes burst through the boiling pot with reckless fury, flinging lumps of mud into the air and producing a fascinating array of colors. The dark slime generally consists of volcanic ash churned to mud in underground cavities by the agitated water.

A fumarole is formed when water vapor or hot gases escape from a vent or fissure at temperatures above the boiling point of water. It is considered a minor volcanic form, since most of the gases are generally volcanic in origin, in contrast with hot springs and geysers, which are composed almost entirely of recirculated ground water. Some of the gases are foul-smelling. If the gases are mainly sulfurous, the formation is called a *solfatara*. The mouth of a fumarole may be decorated with sparkling crystals in various colors, deposited by the mineral-rich vapors as they condense.

On August 8, 1980, *The New York Times*, in an article on the national-park system, noted that private interests were seeking to develop geothermal energy sources just outside the boundaries of Yellowstone National Park despite fears that tapping the underground heat and steam power would change the pattern that sustains the spectacular thermal phenomena inside the park.

Inexpensive and pollution-free geothermal waters supply the heat for thousands of Icelandic homes and for many hothouses that produce fruits, vegetables, and flowers. One of the pioneer plants for the utilization of geothermal energy is in the Larderello area south of Florence, Italy. The superheated steam extracted from the wells produces large quantities of electric power at a fraction of the cost of hydroelectric energy. The plant also supports a thriving chemical industry. In New Zealand only restricted areas have been opened for exploitation, and these produce more than 10 percent of the country's electrical needs. The first commercial geothermal station in the United States has been established at Geyserville in California, ninety-five miles north of San Francisco. The quantity of energy locked up in the volcanic interior of the earth is incalculable. This may be the source the industrial society may have to exploit in the foreseeable future, if the hazards and limitations of fossil fuel are to be overcome. But to extract a meaningful supply of energy for worldwide use, the techniques of exploitation will have to go beyond those few areas where hot springs and geysers spout from the earth.

GUIDE TO COMMON EVERGREENS OF THE MIDDLE ROCKY MOUNTAINS

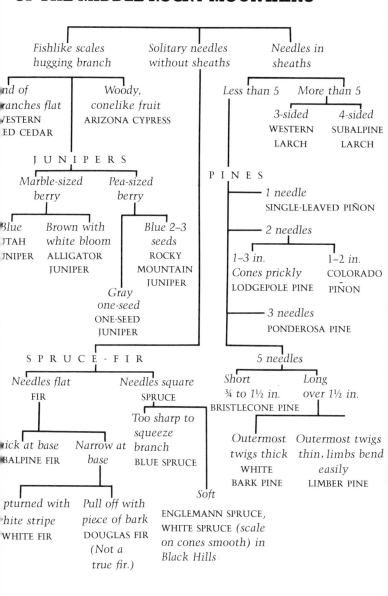

Fishlike scales hugging branch

Solitary needles without sheaths

Needles in sheaths

nd of ranches flat
WESTERN ED CEDAR

Woody, conelike fruit
ARIZONA CYPRESS

Less than 5

More than 5

3-sided
WESTERN LARCH

4-sided
SUBALPINE LARCH

JUNIPERS

Marble-sized berry

Pea-sized berry

PINES

1 needle
SINGLE-LEAVED PIÑON

Blue
UTAH UNIPER

Brown with white bloom
ALLIGATOR JUNIPER

Blue 2–3 seeds
ROCKY MOUNTAIN JUNIPER

2 needles

1–3 in. Cones prickly
LODGEPOLE PINE

1–2 in.
COLORADO PIÑON

Gray one-seed
ONE-SEED JUNIPER

3 needles
PONDEROSA PINE

SPRUCE-FIR

5 needles

Needles flat
FIR

Needles square
SPRUCE

Short ¾ to 1½ in.
BRISTLECONE PINE

Long over 1½ in.

Too sharp to squeeze branch
BLUE SPRUCE

Outermost twigs thick
WHITE BARK PINE

Outermost twigs thin, limbs bend easily
LIMBER PINE

ick at base
BALPINE FIR

Narrow at base

pturned with hite stripe
WHITE FIR

Pull off with piece of bark
DOUGLAS FIR
(*Not a true fir.*)

Soft
ENGLEMANN SPRUCE, WHITE SPRUCE (*scale on cones smooth*) in *Black Hills*

COMMON JUNIPER

1. *Not a tree but a shrub.*
2. *Awl-shaped "leaves" in whorls of 3, prickly, standing out from stem.*
3. *Fruit dark blue, juicy, pea-shaped but small, used to flavor gin.*

SIMPLIFIED GUIDE TO MIDDLE ROCKY MOUNTAIN TREES WITH LEAVES

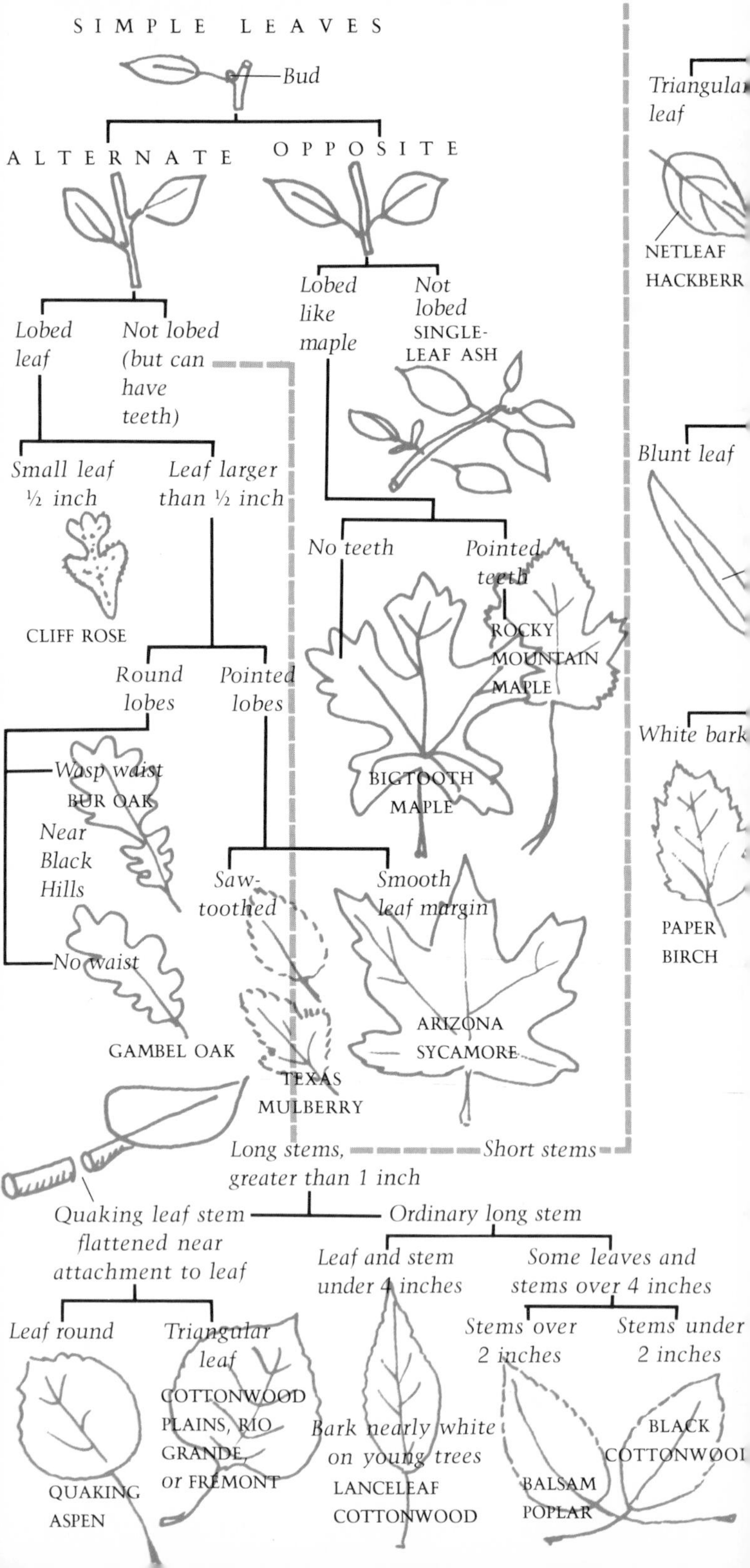

SIMPLE LEAVES

Bud

ALTERNATE — OPPOSITE

ALTERNATE

Lobed leaf — Not lobed (but can have teeth)

Small leaf ½ inch — Leaf larger than ½ inch

CLIFF ROSE

Round lobes — Pointed lobes

Wasp waist
BUR OAK

Near Black Hills

No waist

GAMBEL OAK

Saw-toothed

TEXAS MULBERRY

OPPOSITE

Lobed like maple — Not lobed
SINGLE-LEAF ASH

No teeth — Pointed teeth
ROCKY MOUNTAIN MAPLE

BIGTOOTH MAPLE

Smooth leaf margin

ARIZONA SYCAMORE

Triangular leaf

NETLEAF HACKBERRY

Blunt leaf

White bark

PAPER BIRCH

Long stems, greater than 1 inch — Short stems

Quaking leaf stem flattened near attachment to leaf — Ordinary long stem

Leaf round — Triangular leaf
COTTONWOOD PLAINS, RIO GRANDE, or FRÉMONT

QUAKING ASPEN

Leaf and stem under 4 inches — Some leaves and stems over 4 inches

Bark nearly white on young trees
LANCELEAF COTTONWOOD

Stems over 2 inches — Stems under 2 inches

BALSAM POPLAR

BLACK COTTONWOOD

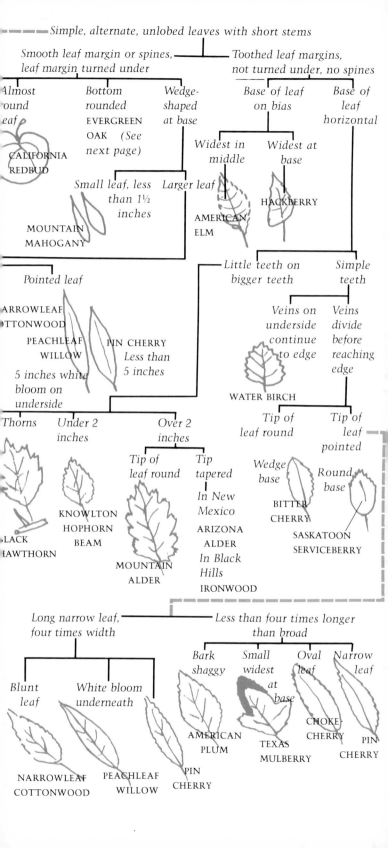

Simple, alternate, unlobed leaves with short stems

Smooth leaf margin or spines,
leaf margin turned under

Toothed leaf margins,
not turned under, no spines

Almost
round
leaf

CALIFORNIA
REDBUD

Bottom
rounded
EVERGREEN
OAK (See
next page)

Wedge-
shaped
at base

Base of leaf
on bias

Base of
leaf
horizontal

Small leaf, less
than 1½
inches

Larger leaf

Widest in
middle

Widest at
base

HACKBERRY

MOUNTAIN
MAHOGANY

AMERICAN
ELM

Pointed leaf

ARROWLEAF
COTTONWOOD

PEACHLEAF
WILLOW

5 inches white
bloom on
underside

PIN CHERRY
Less than
5 inches

Little teeth on
bigger teeth

Simple
teeth

Veins on
underside
continue
to edge

Veins
divide
before
reaching
edge

WATER BIRCH

Thorns

Under 2
inches

Over 2
inches

Tip of
leaf round

Tip of
leaf
pointed

Tip of
leaf round

Tip
tapered

In New
Mexico

ARIZONA
ALDER

In Black
Hills

IRONWOOD

Wedge
base

BITTER
CHERRY

Round
base

SASKATOON
SERVICEBERRY

KNOWLTON
HOPHORN
BEAM

BLACK
HAWTHORN

MOUNTAIN
ALDER

Long narrow leaf,
four times width

Less than four times longer
than broad

Blunt
leaf

White bloom
underneath

Bark
shaggy

Small
widest
at
base

Oval
leaf

Narrow
leaf

AMERICAN
PLUM

PIN
CHERRY

TEXAS
MULBERRY

CHOKE-
CHERRY

PIN
CHERRY

NARROWLEAF
COTTONWOOD

PEACHLEAF
WILLOW

SIMPLIFIED GUIDE TO MIDDLE ROCKY MOUNTAIN TREES WITH LEAVES (cont'd)

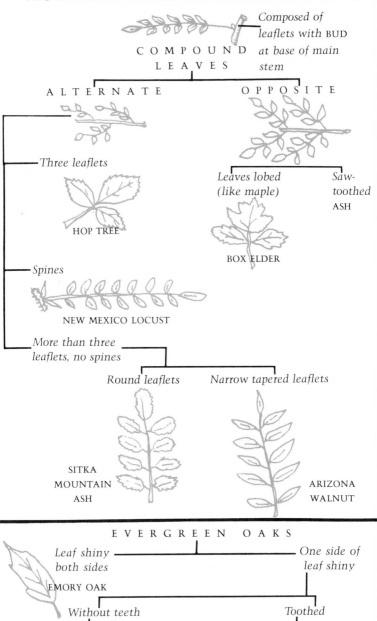

COMPOUND LEAVES — *Composed of leaflets with* BUD *at base of main stem*

ALTERNATE — OPPOSITE

Three leaflets — HOP TREE

Leaves lobed (like maple) — BOX ELDER

Saw-toothed ASH

Spines — NEW MEXICO LOCUST

More than three leaflets, no spines

Round leaflets — SITKA MOUNTAIN ASH

Narrow tapered leaflets — ARIZONA WALNUT

EVERGREEN OAKS

Leaf shiny both sides — EMORY OAK

One side of leaf shiny

Without teeth

Tip pointed — SILVERLEAF OAK

Tip rounded — MEXICAN BLUE OAK

Toothed

Dull blue-green, hairy underneath — ARIZONA WHITE OAK

Hollylike leaves, whitish underneath — CANYON LIVE OAK

Selected National Parks and Monuments

The West is particularly rich in national parks and monuments. These have been set aside for the enrichment of the lives of all the American people. They preserve areas of specific scenic, historic, and scientific value. The vacationer seeking to refresh his spirit has many to choose from. Those described below have outstanding general interest. They are listed in alphabetical order.

Arches National Park. The largest concentration of natural stone arches in the world is located in southeastern Utah. They range from holes as small as house windows to a tall, straddling arch on the rim of a large amphitheater. One arch 291 feet long is thought to be the world's longest natural span; it seems to stand in defiance of all architectural principles of stability. The geological story begins with the upward warping of the earth's crust, followed by the cracking of a 300-foot layer of sandstone into vertical sections. Over millions of years, most of these sections, or fins, were eroded away, but some have remained in the forms of these breathtaking marvels of nature. There are many small birds, squirrels, kangaroo rats, rabbits, and reptiles. At night, deer, coyotes, and foxes move about. During the summer, the moist valleys are covered with wildflowers. Most of the area is high and dry, covered with scattered growths of yuccas, cacti, piñon pine, and juniper.

Badlands National Monument. Some sixty miles east of Rapid City, South Dakota, is one of the most rugged examples of a landscape in the process of being torn away by erosion. Wind, rain, heat, and frost keep carving sharp ridges, gullies, pyramids, pinnacles, and steep-walled canyons out of the soft sedimentary layers of earth, revealing the bones of mammals that inhabited the region twenty-five million years ago. It is raw and arid, supporting little life, except in the wet areas, where the plants and animals are similar to those seen on the neighboring prairies. Small animals inhabit the green islands of cottonwoods, and here and there is a prairie dog town, together with the badgers and coyotes that feed on its inhabitants. The cliffs provide nesting sites for swifts, swallows, and wrens. Junipers and yuccas become established on the slopes and in the valleys, where there are mice, chipmunks, jackrabbits, cottontails, and a variety of snakes. Deer and pronghorn have returned, and bison and bighorn have been reintroduced to restore some of the spectacle of the last century.

Black Hills. This isolated group of mountains in western South Dakota, covering an area of some six thousand square miles, is covered by thick forests of deep-green pines and spruce that seem black in contrast to the surrounding landscape of farm- and ranchlands. It is an area rich in legends of the Old West, and includes such attractions as Mount Rushmore National Memorial (with Gutzon Borglum's massive sixty-foot-high tribute to four American presidents), Wind Cave National Park, Custer State Park, Jewel Cave National Monument, Spearfish Canyon, Deadwood Gulch, Crazy Horse Monument, and Black Hills National Forest. Situated near the geographical center of the United States, the Black Hills is a natural crossroads for the flora and fauna of the major surrounding ecological systems. Deciduous trees from the east, pines from the west, spruce from the north, even yucca and cactus from the south, and the diverse plant and animal life associated with each of these communities meet and overlap to create a naturalist's paradise.

Bryce Canyon National Park. This area, not really a canyon, encompasses sixty square miles of the eroded eastern face of the Paunsaugunt Plateau in southern Utah. The sculptural forms resulting from the erosion of relatively soft layers of rock are among the most colorful in the world. Cathedrals, castles, pinnacles, towers, and domes in shades of pink, yellow, orange, and red, tempered by grays and browns, all create a fanciful wonderland that challenges the imagination. The Native Americans described the formations as "red rocks standing like men in a bowl-shaped canyon." Where the incline is not too steep, separate communities of plants and animals become established at different elevations. Along the lower hills, there is a pygmy forest of piñon pine, juniper, Gambel oak, and sagebrush. Ponderosa pine, Rocky Mountain juniper, and bitterbush cover the upper slopes. At the higher elevations, the dominant community is the spruce-fir forest. Mule deer browse in the meadows. Chipmunks and golden-mantled ground squirrels are plentiful. Other animals in the forests include the yellow-bellied marmot, badger, fox, and porcupine.

Canyonlands National Park. Spectacular canyons have been cut into the earth in this part of southeastern Utah to a depth of 1,500 feet by the Colorado and Green rivers. These rivers converge in the park to provide one of the wildest white-water rides in the country, a ride that ends in the quiet waters of man-made Lake Powell. The entire region is filled with diverse rock formations. Apart from the canyons, there are plateaus, buttes, mesas, spires, arches, needles, domes, castles, standing rocks, sculptured fins, and other fantastic shapes and forms. The country is hot and dry, with wildlife restricted to desert-zone animals. Thunderstorms in late summer produce raging torrents that tear through small gorges, continuously changing the shape of the landscape. There is much fossil evidence of the park's earliest inhabitants.

Capitol Reef National Park. The term "reef" is used to describe a land formation that acts as a barrier to travel. The reef in this park in central Utah was named for a high point that resembles the fluted dome of the U.S. Capitol. The area encompasses a tilt in the earth's crust that exposes layers of rock that have eroded into cliff faces, arches, towers, domes, and pinnacles in a variety of vivid colors. The entire setting is in a desertlike environment where the average annual rainfall is less than seven inches. There is lush growth along the river bottom, but on the exposed dry slopes and terraces, plants struggle to hold on against the sweep of a flash flood. Pockets of water remain after a shower only long enough for tadpoles to go rapidly through their life cycles. Before the water evaporates, they become toads, which survive until the next rainfall by burying themselves in the sand.

Cedar Breaks National Monument. In southern Utah's high country, a huge, many-colored amphitheater is still being slowly scooped out of the pink cliffs of the plateau by the forces of nature. From the road on the rim of the canyon, which passes through forests and meadows, there are panoramic views of mineral-stained spires and ridges. The name "Cedar Breaks" comes from the word "breaks"—used by the early settlers to describe badlands—and from their erroneous use of "cedar" to identify the junipers that grow near the base of the cliffs. On the rim, more than ten thousand feet above sea level, the subalpine environment is filled with majestic stands of pine, spruce, and quaking aspen, with scattered meadows of alpine vegetation above the treeline. There are also small stands of venerable bristlecone pine, some more than a thousand years old. Mule deer graze in the meadows along the rim, squirrels and chipmunks are busy in the pine and spruce forests, and pikas and marmots make their homes among the rocks on the high slopes. Among the many birds, Clark's nutcracker is easily identified around the campgrounds, while the violet-green swallow and the white-throated swift seem to fly endlessly just below the rim.

Death Valley National Monument. In the western part of the Mohave Desert, just inside California on the Nevada border, is Death Valley, the lowest and hottest spot in the United States. Hemmed in on all sides by mountain ranges is an incredible showcase of geological wonders, with salt-water springs, dry lakes, sand dunes, borax flats, and the "badland" features of wind and water erosion. Except in the salt flats, there is a varied and abundant plant and animal life. At higher elevations, there are forests of junipers, mountain-mahogany, and piñon pine. In the lower regions, there is a wide spectrum of desert life, with birds, lizards, snakes, and nocturnal rodents, and in a favorable springtime, a vivid display of desert wildflowers. The valley is unbearably hot in the summer. Visits are best made between October and May.

Grand Canyon National Park. Part of one of the most spectacular scenic wonders of the world, situated in northwestern Arizona, has been preserved in this park. The colors of this awesome spectacle constantly change from dawn to sunset. The mile-deep canyon, carved by the Colorado River over more than two hundred miles, exposes, layer by layer, an amazingly revealing record of the eras of geological time. Climatic conditions at the bottom of the canyon are almost desertlike; those on the North Rim approximate a cool mountain environment. Life zones range from the near-tropical to the sub-arctic. Cacti, agaves, and mesquite are found in the lowest elevations, and pine, spruce, fir, and aspen in the highest areas. The canyon acts as a barrier to the movement of wildlife. The squirrel in the ponderosa-pine forest on the north has a black belly and pure white tail, while its relative on the south side has a white belly and a tail that is whitish only on the undersides. Separated for generation after generation, these descendants of a common ancestor have developed these differences in coloration. It is difficult to overstate the majestic wonder of this grand canyon.

Grand Teton National Park. Immediately south of Yellowstone National Park, the blue-gray pyramids of the Grand Tetons soar more than a mile above the meadows and lakes of the Jackson Hole basin. All the major canyons show evidence of the profound sculpturing accomplished by the glaciers of the Ice Age. The glaciers now visible along the steep eastern front are pygmies compared to those of the past, but they are still quarrying away into the face of the mountains. Rising from the valley floor are forests of lodgepole pine, Engelmann spruce, Douglas fir, and limber and whitebark pine. Along the streams are cottonwoods, as well as the aspens and willows used by the beavers. Elk and mule deer range from the valley to the timberline, and moose feed on aquatic vegetation. At higher elevations, the tiny pika and the yellow-bellied marmot share the barren rock habitats. Lower down, chipmunks and squirrels are plentiful. There are floral displays at all elevations, and hundreds of bird species—from the tiny hummingbird to the soaring eagle and, sometimes, even the rare trumpeter swan.

Great Sand Dunes National Monument. The dunes in southern Colorado are made up of volcanic rock fragments and bits of quartz from the components of the mountains that surround the San Luis Valley. For thousands of years, winds from the southwest have driven these sands into a pocket created by the flanks of the fourteen-thousand-foot peaks of the Sangre de Cristo Mountains, creating more than fifty square miles of sand dunes that sometimes reach seven hundred feet high. To escape the heat and dryness of the dunes, much of the animal life is subterranean and nocturnal. The extent of each night's activity can be seen the following morning in the variety of tracks and markings left in the sand. In the valley floor surrounding the dunes, there are many birds and small mammals among the sagebrush, junipers, and cottonwoods. Most of the plant life begins with small patches of grass, peas, and sunflowers that become established in the protected flats between the dunes. In the rabbitbrush and grassland of the valley floor, there are small mammals and birds. The piñon-juniper belt of the foothill region abounds with chipmunks, rabbits, ground squirrels, mule deer, coyotes, magpies, and jays.

Joshua Tree National Monument. In southern California, where the "high" Mohave Desert meets that portion of the Sonoran Desert known as the "low" Colorado Desert, 872 square miles have been set aside to protect a rich desert vegetation that includes the giant yucca tree that was given the name "Joshua" by traveling Mormons. Beginning at the headquarters at Twentynine Palms, the road through the monument affords a view of an astonishing variety of plants and animals. Apart from the spectacular Joshua trees, found nowhere outside the Mohave Desert, there are Mohave yuccas, Spanish bayonets, cholla cacti, and a number of shrub communities. All these harbor a considerable amount of animal activity, including desert tortoises, ground squirrels, and crested lizards, as well as roadrunners, thrashers, and cactus wrens. Of special interest are the natural oases surrounded by palms and other water-tolerant plants. These protected pools attract many of the birds and some of the larger animals.

Mesa Verde National Park. Among the best examples of cliff dwellings in the United States are those preserved in the rock shelters of Mesa Verde in southwestern Colorado. These dwellings were built in large crevices that had been carved by wind and water erosion into the high cliffs above the canyon floors of this mesa. The inhabitants left their homes on the top of the mesa to move into these inaccessible and more defensible shelters, although they continued to farm the fields above them. By the close of the thirteenth century, these cliffs were abandoned. The present fragile condition of the sites that have thus far been uncovered and preserved and the historical value of those still buried have resulted in the imposition of limitations on tourist travel. To a large extent, the flora and fauna of the mesa remain as they were when the Indians left it. The plateau is covered with piñon pine, juniper, oak, and ponderosa pine, with hardy shrubs in the drier areas of the canyons.

Petrified Forest National Park. In northeastern Arizona, about 250 miles from Phoenix, the Petrified Forest and Painted Desert combine to provide a vast horizon of weird, eroded badlands colored in purple, red, yellow, and browns, and the remnants of a pine forest, 150 to 200 million years old, that has turned to stone. Scattered throughout are thousands of great logs sparkling with jasper, carnelian, and agate. These logs reveal the smallest details of the original structure of their woody tissues, preserved intact as a result of the process of petrification. In the Painted Desert, the vivid colors in the various layers of clay in the mesas, buttes, and plateaus constantly change during the day. Because the region receives less than ten inches of rain a year, hardy desert flowers, such as cacti, yucca, and rabbitbrush, are dominant. Despite the scant vegetation, many animals—including cottontails, jackrabbits, coyotes, and pronghorn—survive in the region. The park contains ruins of Indian dwellings and drawings and signs (petroglyphs) etched in the surface of stone by the early inhabitants.

Rocky Mountain National Park. Starting some sixty-five miles from Denver is a tract of four hundred square miles that provide a thrilling introduction to the heart of the southern Rockies. Accessibility by automobile has been simplified by Trail Ridge Road, which reveals the pageantry of these mountains as it winds for fifty miles to an elevation of more than twelve thousand feet and for eleven miles remains above treeline. Plant and animal life change dramatically in the trip from the valleys to the high peaks. At lower altitudes, stands of juniper and ponderosa pine face the sun; on the most shady northern slopes, Douglas fir will appear and finally take over in the higher elevations. Thickets of quaking aspen stand out in delicate relief. In the highest forest zone, spruce and fir are the dominant species. At the timberline, fierce winds twist and reduce the trees into grotesque and ground-hugging specimens. Then the trees disappear, and spread out in every direction is the land of the tundra. At every elevation there are different spectacles of wildflowers and occasional glimpses of a variety of animal life in its natural habitat. Views of snow-covered peaks, pebble-bottomed streams, stream-cut valleys, and glacier-carved lakes are everywhere.

Saguaro National Monument. This monument, outside of Phoenix, is named for the majestic saguaro, a plant with many thorny arms, which can grow fifty feet tall and endure heat, cold, cloudburst, and drought for 150 years. The saguaro itself is the center of considerable animal activity: doves, bats, and insects feed on the nectar of its blossoms; its fruit is food for birds; its stems are nesting places for woodpeckers, flickers, owls, and flycatchers; and its roots and branches provide shelter for snakes and rodents. The setting is a spectacular desert with tree-sized mesquite, ocotillo, paloverde, and ironwood plants, with jointed-stem, cylinder, and ribbed cacti, with spectacular springtime displays of annual wildflowers, with the Gila monster, chuckwalla, crested lizard, horned lizard, and desert tortoise, with the remarkable kangaroo rat, and with mule deer, bighorn sheep, and coyote. The famed Arizona-Sonoran Desert Museum provides a fascinating closer look at much of the plant and animal life that flourishes in the surrounding desert.

White Sands National Monument. The sands of the largest gypsum desert in the world, located in south-central New Mexico, come from the sedimentary layers of rock in the mountains that imprison the Tularosa Basin. Among the layers of rock are formations of gypsum that dissolve in rainwater and snowmelt, run off into the basin, and evaporate into crystals, which are then pulverized into fine grains and whirled across the desert into gleaming dunes. Yuccas, sumacs, and cottonwoods have developed special adaptations to deal with the threat of suffocations from advancing dunes. Many of them do not survive and, when dunes move on, leave behind pedestals of gypsum held together by their networks of exposed roots. Only a handful of small creatures has adapted to this challenging environment. Some, like the earless lizard and the Apache pocket mouse, have each evolved into a race of bleached specimens that blend with the snow-white dunes. Some, like the rabbits, survive because of their enormous rate of reproduction.

Wind Cave National Park. Preserved in the southwestern section of the Black Hills is a distinctive type of limestone cavern, named for the strong currents of air that blow alternately in and out of the cave—a phenomenon apparently caused by the rise and fall in atmospheric pressure on the outside. Some of the subterranean rooms and passages are decorated with crystalline formations in unique honeycomb patterns called "boxwork." Rivaling this attraction is the surrounding wildlife sanctuary that preserves, over more than forty square miles, a prime example of a mixed-grass prairie, with tall, medium, and short grasses that sparkle with a large assortment of wildflowers in spring and summertime. Grazing in these lush open spaces are many species of animals native to the Old West. There are herds of bison, as well as pronghorn, deer, and elk. There are several towns of black-tailed prairie dogs, and badgers, raccoons, and coyotes. Birds are abundant and, as with plant life, include inhabitants of the diverse geographic environments that meet and overlap in the Black Hills. Forests of ponderosa pine rise above the rolling plains.

Yellowstone National Park. Situated mainly in northwestern Wyoming, this is the oldest (established in 1872) and largest (more than 3,400 square miles) of the national parks, encompassing a wonderland of thermal phenomena, a large variety of scenic splendors, and an unusually large concentration of native wildlife. There are more than 10,000 hot springs, mud pots, steam vents, boiling pools, and spectacular geysers hurling tons of water into the sky, with Old Faithful establishing a record for the promptness of its performance. Ridges and mountains surround the central plateau, the Yellowstone River cuts through a canyon a thousand feet deep where ospreys wheel and soar continuously, the upper and lower falls break into roaring white jets, and forests of lodgepole pine and other evergreens fill the landscape. The wildlife is unsurpassed. Bighorn sheep climb the crags, elk browse in meadows of wildflowers, wary pronghorn survey the hillsides, moose wade into rivers and lakes, bears roam through the forested areas and onto the open highways, and, in the winter, buffalo forage around the snow-free geysers. The whole park is visible from the windswept summit of Mt. Washburn.

Zion National Park. Different stages in the geological story of the natural processes that have been operating for billions of years are revealed in three of the national parks. The most ancient rock layers can be seen in the depths of the Grand Canyon. The most recent, and uppermost, layers make up the cliffs in Bryce Canyon, where forces of erosion have been less effective. The middle period of earth's history is told in the amber-colored sandstones of Zion Canyon in southwestern Utah. Here, the erosional forces have opened up for inspection the era of the dinosaurs amid rock formations streaked with brilliant colors. Against the background of the stone walls of the canyon, there are diverse environments of plants and animals: a scant cover of cacti and yucca in the low desertlike region; growths of broadleaf trees along the river at the base of the canyon; piñon-juniper forests on the drier open areas; low shrubs in the open flats spotted with thickets of short evergreens and oaks; ferns and hanging gardens of water-loving flowers around many springs; and fir, pine, and aspen in the high country. Many birds and animals migrate from the canyon to the higher elevations in the summer, then back when the snow begins to fall.

A SELECTED LISTING OF PARKS AND MONUMENTS, STATE BY STATE

The following is a briefly annotated, selected listing of places, alphabetically by state. Many of the more popular areas have associated with them a non-profit association that publishes and distributes books and pamphlets about the history, geology, and natural features of the area. A price list can be obtained by writing to the superintendent. Our experience has suggested that this be done six months before you intend to visit. Although these pamphlets can often be obtained on the site, a better selection is available ahead of time; in peak tourist season, supplies often run out.

The following abbreviations have been used:

NM	=	National Monument
NMem	=	National Memorial
NRA	=	National Recreation Area
NHS	=	National Historic Site
NWR	=	National Wildlife Refuge
NMemP	=	National Memorial Park
NP	=	National Park

Arizona

Canyon de Chelly (NM), Box 588, Chinle, AZ 86503.
  A popular Navajo Indian post; entry only with guided tour.
Casa Grande Ruins (NM), Box 518, Coolidge, AZ 85228.
  A four-story tower with walls four feet thick of poured mud.
Chiricahua (NM), Dos Cabezas Star Route, Willcox AZ 85643.
  A desert upland with Mexican species.
Coronado (NMem), Star Route, Hereford, AZ 85615.
Glen Canyon (NRA), Box 1507, Page, AZ 86040.
  Mostly unvegetated "slick rock"; a major water-sports area.
Grand Canyon (NM), c/o Grand Canyon NP.
Grand Canyon (NP), Box 129, Grand Canyon, AZ 86023.
Lake Mead (NRA), 601 Nevada Highway, Boulder City, Nevada 89005.
  Also in Nevada. Contains one hundred miles of the Grand Canyon.
Montezuma Castle (NM), Box 219, Camp Verde, AZ 86322.
  Neither Aztec nor castle but a cliff dwelling.
Navajo (NM), Tonalea, AZ 86044.
  Cliff dwelling.
Organ Pipe Cactus (NM), Box 100, Ajo, AZ 85321.
  A wild desert lowland, popular in winter.
Pipe Spring (NM), c/o Zion NP, Springdale, Utah 84767.
Petrified Forest (NP), Holbrook, AZ 86025.
  A desert upland with the famous petrified wood.
Saguaro (NM), Box 17210, Tucson, AZ 85710.
  A desert lowland from 2,700 feet to 9,000 feet, many ecosystems.
Sunset Crater (NM), c/o Wupatki NM, Tuba Star Route, Flagstaff AZ 86001.
  Volcano erupted in 1064; crater still visible.

Tonto (NM), Box 1088, Roosevelt, AZ 85545.
  Cliff dwellings.
Tumacacori (NM), Box 67, Tumacacori, AZ 85640.
  Pronounced "Too-mah-COCK-oh-ree"; an ancient mission.
Tuzigoot (NM), Box 68, Clarkdale, AZ 86324.
  A pueblo that flourished for three centuries.
Walnut Canyon (NM), Route 1, Box 790, Flagstaff, AZ 86001.
  Although they were established as a monument in 1915, pot hunters have
  done a lot of damage to these ruins.
Wupatki (NM), Tuba Star Route, Flagstaff, AZ 86001.
  One of the larger ancient pueblos.

California
Joshua Tree (NM), Box 875, Twentynine Palms, CA 92277.
  A well-liked lowland desert area.
Death Valley (NM), Death Valley, CA 92328.
  282 feet below sea level, a record temperature of 134°F.
Anza-Borrego State Park, Borrego Springs, CA 92004.
  Vast and desolate, spring wildflowers; half-million visitors a year.
Providence Mountains State Recreational Area, Box 1, Essex, CA 92332.
  In the heart of the Mohave Desert but high enough to be relatively cool.

Colorado
Black Canyon of the Gunnison (NM), c/o Curecanti NRA, Montrose, CO
  81401.
  Impressive look down.
Colorado (NM), Box 438, Fruita, CO 81521.
  Mesa and canyon environment, scenic road.
Dinosaur (NM), Box 201, Dinosaur, CO 81610.
  Concentration of dinosaur bones and beautiful scenery. Also in Utah.
Great Sand Dunes (NM), Box 60, Alamosa, CO 81101.
  Some of the world's tallest dunes, 700 feet high.
Hovenweep (NM), c/o Mesa Verde NP, CO 81330.
  Ruins, also in Utah.
Mesa Verde (NP), Mesa Verde NP, CO 81330.
  Ancient cliff dwellings, excellent example of oak glens.
Rocky Mountain (NP), Box 1080, Estes Park, CO 80517.
  Good road into tundra, other high environments.
Shadow Mountain (NRA), c/o Rocky Mountain NP, Box 1080, Estes Park,
  CO 80517.
  Area on western side of Rocky Mountain NP.

Idaho
Craters of the Moon (NM), Box 29, Arco, ID 83213.
  Volcanic landscape.
Yellowstone (NP), Yellowstone NP, *Wyoming* 82109.
  The first and greatest national park. Also in Wyoming, Montana.

Montana
Glacier (NP), West Glacier, MT 59936.
  North of our area, known for its hiking trails. Also extends into Canada.
Yellowstone (NP), Yellowstone NP, *Wyoming* 83020.
  Also in Idaho, Wyoming.
Bighorn Canyon (NRA), Box 458, Hardin, MT 59035.
  Introduced wild horses, centers around Bighorn Lake. Also in Wyoming.

Red Rock Lakes (NWR), Monida Star Route, Lima, MT 59739.
  Famous for its trumpeter swans; 50 birds left in 1935 have increased to
  hundreds.

New Mexico
Bandelier (NM), Los Alamos, NM 87544.
  Most of the monument is wilderness; features accessible only by trail.
Carlsbad Caverns (NP), Capulin, NM 88414.
  Chihuahuan Desert flora as well as the celebrated caves.
White Sands (NM), Box 458, Alamogordo, NM 88310.
  Pure gypsum dunes over 176,000 acres.

South Dakota
Badlands (NM), Box 72, Interior, SD 57750.
  Grasslands and arid areas.
Mt. Rushmore (NMem), Keystone, SD 57751.
  The famous presidential sculptures.
Wind Cave (NP), Hot Springs, SD 57747.
  Grasslands with bison herd and caves.
Custer State Park, Hermosa, SD 57744.
  In the Black Hills.

Texas
Big Bend (NP), Big Bend, TX 79834.
  Chihuahuan Desert plants unfamiliar to most Americans.
Guadalupe Mountains (NP), 3225 National Parks Highway, Carlsbad, *New
  Mexico* 88220.
  A new, relatively undeveloped park.

Utah
Arches (NM), c/o Canyonlands NP, Uranium Bldg, Moab, UT 84532.
  Largest concentration of natural arches in the world.
Bryce Canyon (NP), Bryce Canyon, UT 84717.
  A favorite little jewel. One looks down on fantastic scenery.
Canyonlands (NP), Uranium Bldg, Moab, UT 84532.
  Vast area of canyons, arches, colored cliffs, ruins.
Capitol Reef (NM), Torrey, UT 84775.
  Rock domes that look like capitol buildings, red cliffs, canyons.
Cedar Breaks (NM), c/o Zion NP, Springdale, UT 84767.
Timpanogos Cave (NM), RFD 3 Box 200, American Fork, UT 84003.
  A walk up and then a cave.
Zion (NP), Springdale, UT 84767.
  A canyon cut by the Virgin River from which one looks up at fantastic
  scenery.

Wyoming
Devils Tower (NM), Devils Tower, WY 82714.
  The old plug of a volcano standing alone.
Grand Teton (NP), Box 67, Moose, WY 83012.
  Most like the Alps.
Yellowstone (NP), Yellowstone NP, WY 82109.
  The first and greatest national park. Also in Montana and Idaho.

Index

(Note: Page numbers in boldface indicate illustrations.)

952